The Continental Shelves

The Continental Shelves

By John F. Waters

ABELARD-SCHUMAN
NEW YORK

 Published simultaneously in Canada by Fitzhenry & Whiteside Limited, Toronto. Manufactured in the United States of America.

Library of Congress Cataloging in Publication Data

Waters, John Frederick, 1930–
The continental shelves.
Bibliography: p. Includes index.
Summary: Describes the continental shelves and discusses their present and potential uses. 1. Continental shelf—Juvenile literature. [1. Continental shelf] I. Title. GC85.W37 1975 551.4'1 75–6697 ISBN 0–200–00157–4

Contents

Illustrations

Acknowledgments

The author would like to thank the following persons for their help in preparing this book: Bruce Crawford, Dr. Gifford C. Ewing, Dr. Robert L. Guillard, George R. Hampson, Dr. Robert R. Hessler, Tai Soo Park, Dr. Howard L. Sanders, Dr. Rudolf S. Scheltema, John L. Schilling, Thomas R. Stetson, Harry J. Turner, Dr. David Wall, Richard L. Haedrich, Nat Corwin, George D. Grice of the Woods Hole Oceanographic Institution. Also, F. W. McCoy of Harvard University; Paul Ferris Smith of Woods Hole; the late Dr. C. Lloyd Claff of the Marine Biological Laboratory, Woods Hole; and special appreciation to Dr. K. O. Emery, Dr. Mary Sears, Charles S. Innis, David M. Owen, and Dr. David A. Ross of the Woods Hole Oceanographic Institution.

The author would also like to thank the following institutions and companies for their help: University of Rhode Island; Ocean Systems Inc.; Birns & Sawyer, Inc.; The University Museum of the University of Pennsylvania;

Rebikoff Underwater Products, Inc.; Reynolds Metals Company; Massa Division of the Dynamics Corporation of America; NUS Corporation, Underwater Systems Division; Westinghouse Electric Corporation, Underseas Division; Marine Acoustical Services of Miami; Reynolds Sub/Marine Services Corporation; Global Marine Inc.; Scripps Institution of Oceanography; Bendix Electrodynamics Division; The Bissett-Berman Corporation; General Dynamics Corporation, Electric Boat Division; Marineland of the Pacific; National Science Foundation; Miami Seaquarium; United States Navy Oceanographic Office; National Marine Fisheries Service, Woods Hole; American Petroleum Institute; Exxon Corporation; Coastal Services, Division of Ocean World Corporation.

The Continental Shelves

1 Underwater Platforms

Just about everyone knows what the ocean is. Even if they have never seen it close up, they know that it is a vast area of water, salty water, that covers almost three-fourths of the earth's surface. They also know that where the ocean meets the land is an area usually called a shoreline. And some of these shorelines are in the form of beaches. But when someone asks what is underneath the water, away from the beach, then few people know. They might say that it is the ocean bottom and let it go at that. And they would be right, up to a point. But beyond the beach and underneath the blue-green ocean water is a land of mystery. It is mysterious, and relatively unknown, because this land is always underwater, making it very difficult to study. But men and women who study the ocean and the land beneath it are beginning to learn a great deal about this land that is commonly called the continental shelves.

Anyone who has been to an ocean beach and stood in the water on the sandy or rocky bottom has actually stood on the beginning of the submerged continental shelf. These shelves extend out from most of the continents and some of the large islands all around the world. The shelves vary in width. On mountainous seacoasts, where mountain ranges seem to spring straight up from the sea, the land drops away directly to the depths of the ocean. In these places, such as the western shore of the United States, the shelf is very narrow, sometimes less than five miles. In other regions the shelf stretches out for miles, such as in the Arctic Ocean off the coast of Russia where the shelf extends for 800 miles.

All the shelves of the world, regardless of width, are like giant, tilted underwater platforms. From shore they gradually drop away for 250 to 600 feet on the average, although some drop to 1,700 feet. They meet a steeper slope, called the continental slope. Here the transition from shelf to deep sea is quite sudden. The shelves, averaging a depth of 400 feet, abruptly drop to abysmal depths of 10,000 feet, 20,000 feet and even 30,000 feet, more than five miles down!

Shelves in the underwater world comprise about 10 million square miles, equal to the combined land area of Europe and South America. Yet this relatively modest area is of tremendous importance, so great in fact that without the shelves man would find it very difficult to survive on this planet, if indeed he could survive at all.

Before exploring their importance, it is best to have a better understanding of the makeup of the shelves. They have not always been there nor have they remained the same since they were formed. Exactly how, then, were they formed? Scientists have an idea from what they have been able to learn about the earth, its lands and oceans, through the study of history.

The four great ice ages (Permian, Carboniferous, Cambrian and Precambrian) interest scientists because of their

Model of the Atlantic continental shelf, off the coast of Florida.
PHOTO: U.S. DEPARTMENT OF THE INTERIOR, GEOLOGICAL SURVEY

bearing on the formation of the shelves. During these ice ages much of the earth's water was frozen in thick, shivering ice sheets on the continents. These ice sheets, commonly called glaciers, lowered the level of the sea and exposed the continental shelves. Scientists are sure of this because they have estimated that if the world's present glaciers, giant ice caps in Antarctica and Greenland, were melted, they would raise the present sea level by at least 300 feet and maybe more, depending on who is making the estimate. Several hundred feet may not seem like much, but the raising or lowering of the sea level affects the shorelines of continents by hundreds of miles.

Scientists have studied the shelves throughout the world and they have come to the conclusion that there are rocks underneath them, forming the base. Through the years sediment has built up on the rocks. More than two-thirds of the shelves have been made this way in the past 15,000 years, or since the last lowering of the sea caused by glacial buildup.

One early mystery was why this sediment had not washed away. But scientists learned that the sediment remained in place because of three kinds of underwater dams. One kind, a tectonic dam, was created by cooling volcanic lava. A second, a reef dam, was made by centuries of deposits of marine animals and plants. The third kind, diapir dams, were pushed up from below the ocean floor by salt domes. A few shelves have no dams at all in their makeup and are smooth and sloping. Some of the different kinds of shelves have been further altered by erosion, or changed by giant sheets of ice and endless waves.

Some of this vast submerged land has barrier beaches, elongated ridges of loose sand running parallel to the continents, as well as dams. There are also terraces and sea cliffs. Most shelves do not slope steadily but are terraced. There may be as many as four to six terraces on any given shelf causing it to drop off, slope, drop off and so on. Some terraces are difficult for scientists to find, particularly on steep shelves, where the terraces are very narrow, or on gently sloping shelves, where the terraces are very wide. Also, if the shelves are covered with deep sediment, the terraces underneath remain well hidden.

Some of the world's shelves have been increasing in width and thickness for 100 million years, due to the continuing buildup of sediment. The sediment may come from the mouths of rivers and streams or from the ocean that carries a great deal of sediment by way of ocean currents. And there are shelves, such as the one located along the east coast of the United States, that have no sediment—just a sandy bottom.

Mud is the most common shelf material. Sand is unusual. Sand-covered shelves are usually thinner than those covered with mud and silt. The sandy shelf along the Florida Atlantic shore, for example, is estimated to be almost two miles thick while the mud-bottomed shelf in the Gulf of Mexico, about 100 miles away, is approximately four times as thick.

Scientists have discovered that all along the bottoms or floors of various shelves, in the sand and sediment, are scattered tracings of iron and empty shells of mollusks (like clams or snails). Far out at sea, at the edges of the continental slope, they have found broken shell and shell-sand. Minerals that are usually in the salt water are also in the shelf sediment.

These ocean scientists have also learned that changes occurred on the shelves during the ice ages, when the sea level was lowered. The land was dry and rivers flowed through this land to the new sea level and made their

The shelf bottom at 180 feet, with branch coral and spiny sea urchin at left.

PHOTO: WOODS HOLE OCEANOGRAPHIC INSTITUTION

deposits there. Certain low areas of the newly exposed shelf became quiet ponds and green marshes bursting with life. From that time up to today the debris from meadows and forests has remained in these little basins on the shelf. Freshwater peat has been located in the sea, along the shelf off the coast of the eastern United States and off the coasts of Japan and Europe.

Many of the shelves today have channels that were cut through the terraces by rivers, when the level of the sea was low and the land was dry, and some are quite deep. These crevices are mostly filled with sediment and can only be found with exact oceanographic equipment. It is believed that there are hundreds of these channels on the shelves but only a handful have been located. Therefore a mystery remains as to how many unknown crevices there are today.

Scientists do know there are some large channels, such as the one cut by the Hudson River on the northeast coast of the United States. It leaves New York City and stretches across the shelf until it reaches the continental slope. It is of such size that it has not yet filled with sediment in the thousands of years since it was first formed.

When these channels reach the continental slope, they change into underwater canyons. They end on the sea bottom, sometimes many thousands of feet down. Scientists know that these canyons were not formed by rivers, such as the Hudson channel, because they have always been underwater. It is thought, but not proved conclusively, that the canyons were created by currents that were caused by sediment sliding down the steep slope.

Some of the shelves, such as those on the west coast of the United States, are so narrow that the beginnings of these canyons almost reach the shore. As a result there are sand glaciers. Sand that is brought to shore by erosion of cliffs, the action of the waves or carried by streams is added to the beach for a short time. Then currents move it seaward down through the canyons. The sand behaves like an underwater sand glacier cutting away channels as it moves.

The shelves are a delight to study for geologists, biologists, physicists and chemists because of the many unanswered questions that still exist. One is a definite mystery. What happened to the shelves prior to 100 million years ago? In studies geologists have determined that the shelves have been building for that long, but they do not know what happened before that time.

Very early scientists found out something only if they *had* to know it. About 2,400 years ago, for example, the first ocean scientists made some necessary navigation studies in the Mediterranean Sea. In order to locate Egypt, ancient sailors kept on the lookout for a muddy bottom. When they found it, they knew they were about a day from port.

Before the turn of the twentieth century, there were many soundings taken on the shelves for the purpose of chart making. Also, in 1872, there was the start of the three-and-a-half-year voyage of the research ship *Challenger.* Because of this expedition, and continuing studies of the continental shelf and slope, scientists began to learn about depths, currents, makeup, and the plant and animal life that lived on or above the shelves.

During World War II, study of the shelves increased to improve mine laying, detect sounds at sea and learn tides and currents for amphibious landings. Later, private industry took an interest in the study of the shelves for the sole purpose of seeing if profits could be made there. Governments also turned to the shelves to look for minerals, especially oil. Because of these recent studies the shelves have never been so understood as they are today, but there is still much more to discover.

Why are the shelves so important to man? It is known that the relatively shallow, sun-drenched water above the shelves is teeming with life of all kinds, plants as well as animals. The food chain is greatly in evidence in the sea as well as on land and is dependent on sunlight needed by green plants. Almost all of the organic material, that which feeds the life in the sea, is synthesized in the upper layers of the shelf waters by tiny plants called phytoplankton. These plants, using the sun as a source of energy, make the necessary chemicals to grow. This process is called photosynthesis. Inside each tiny food factory is the chemical chlorophyll, the same that is found in land plants.

The tiny sea plants are eaten by zooplankton, animals usually not larger than the head of a pin. Then slightly larger animals feed on the pinhead-sized animals. Larger animals feed on those that ate the pinhead animals and so on through the food chain until fish such as tuna, swordfish and sharks are feeding on animals of considerable size.

When an animal or plant dies the remains sink and become part of the ooze found on the shelf floor. The bits of animals and plants are then swallowed by such animals as clams, tube worms and sea cucumbers. Bacteria also work on the tissues and break them down into basic chemicals. These chemicals are nutrients. Upward currents, called upwellings, take the nutrients from the shelf floor back up to the sunlit layers, and the chemicals are used again by the phytoplankton, and the food chain continues.

Marine life grows best where the water is cold and rich in dissolved nutrients that flow into the sea from ice-capped seas and rivers. Sometimes the nutrients are a part of an upward flow from the deep ocean. These rich waters are usually over a shelf and may have as many as 12 million chlorophyll plants in a single cubic foot of water. With all this food available, it is correct to assume that there will be a multitude of tiny animals that eat plants and fish that eat tiny animals in the waters over continental shelves.

The world's best fishing areas are where these cold-water pastures are located. One such area is the shelf off the coast of Newfoundland, on the east coast of Canada. Another area is the shelf off the northwest coast of Africa and another along the North Sea, between England and Scandinavia. The fish in these areas are rich in protein because the protein and fat content of a diatom, a tiny one-celled plant, is the same found in rich hay growing in a lush, green meadow on land. There is also a good supply of vitamins A and D, common in fish livers.

The microscopic plants of the shelf waters are not the sole users of the sun's energy. All around the world, along the shores above the shelf, are larger plants such as kelp that reach lengths of 150 feet and more. There is turtle grass and various other kinds of seaweeds. Most of the plants cling to rocks, but a few drift and some are aided by tiny floats filled with gas.

It is also estimated that 90 percent of the food from the ocean, such as fish, clams and oysters, comes from the shelf area and the bays and marshes that connect with the shelf. This valuable food resource is felt to be worth $10 billion each year. Most of the food is in the form of fish.

An area of rapid growth is the natural gas and oil being taken from drilling sites over shelves. In 1970 the value of this drilling was estimated at more than $4 billion and was rising. Of the world's total production of oil this was about one-fifth of the resource taken in 1970. On the shelves

A rapid area of growth is the natural gas and oil being taken from drilling sites over the shelves.
PHOTO: TEXAS GAS TRANSMISSION CORPORATION

encompassing the United States about $1 billion worth of oil was drilled during that same year. It is predicted that the offshore drilling for oil will develop much more rapidly than land-based wells.

There are other materials mined or dredged from the shelves. They include sand and gravel, and some heavy minerals such as iron, tin, diamonds and gold. It is clear that man counts heavily on the shelves for survival and will need to study the shelves and solve their mysteries to discover what more they will offer in the future.

2 The Shelves of Long Ago

Of the discoveries made on or about the shelves, some have a particular interest. These are not the kind of discoveries that are going to determine just how we are going to continue to exist on this planet. They will not decide if our lot is going to be any better or if our life styles will change in the next several generations. These discoveries have value in another sense, that of revealing a bit of history.

In the past two decades rare fossils of animals and plants, considered to be of major scientific importance, were located on the continental shelves off the coasts of Virginia, New York and Massachusetts. These fossils were not discovered as a result of exhaustive scientific research. They were hauled to the surface in the nets of commercial fishermen. The fishermen dragged their nets across the bottom of a great sand area that was laid down about 25,000 years ago when the oceans were several hundred feet below what they are today. The specimens were spotted in the nets by alert men on regular fishing trips and were set aside to be brought into port when the holds of the vessels were filled with the catch.

Once in port, the specimens were shipped to the biological laboratory of the Fish and Wildlife Service (now the National Marine Fisheries Service) in Woods Hole, Massachusetts, and later to the Smithsonian Institution in Washington, D.C.

Some of the fossils brought back were large teeth weighing up to three pounds and measuring eight inches in length. They were identified as the teeth of ice age elephants. These animals, mastodons and mammoths, roamed the shelf from 11,000 to 25,000 years ago. Ice-age mastodons were believed to be similar to specimens that lived in parts of the United States mainland during prehistoric times. But the mammoth was different from mainland forms. Its tooth pattern was similar to the famous woolly mammoth whose frozen remains were found in Alaska and Siberia.

The continental-shelf mammoth may have lived on such coarse plants as pine and spruce needles. Scientists have come to the conclusion that a very large elephant population lived on the continental shelf at the time it was above water. Other bones located have been those of a giant ground sloth, bison, musk oxen and moose. Parts of fish, seals and walrus dating back to the time when the sea started to cover the shelf have also been found. One fish bone resembled the bone of a blowfish or puffer. Compared with modern types, the ancient fish may have been eight feet in length. Similar findings have been made off the coasts of Japan and Europe.

The teeth of mastodons and mammoths have been discovered on the United States east coast continental shelf in more than forty different sites, and in water up to 500 feet deep and 80 miles offshore. Submerged shorelines have also been located, as have relic sands and shells that are commonly found in lagoons.

A four-inch-long fossil shark tooth was brought up from Georges Bank (east of Massachusetts) by the commercial dragger *Explorer* from a depth of 120 feet. The tooth was examined and it was eventually determined that it came from a giant shark that is now extinct. This shark, a close relative of the present-day maneater or great white shark, grew to be 40 to 50 feet long.

A large sample of peat was dredged from the shelf by the *Ruth Lea,* a scalloper out of New Bedford, a seaport in southeastern Massachusetts. Several bushels of peat were collected from a depth of about 200 feet during the dredging operations off Georges Bank. Through studies it was learned that the peat lay on the bottom at the northern end of an underwater sand wave. The peat was found to contain salt-marsh grass, pollen, twigs from spruce, fir and pine trees and spores from peat moss. Some roots of cedar were uncovered as well as a few freshwater one-celled plants, all thriving about 9000 B.C., or 11,000 years ago.

About 13,500 years ago glaciers of enormous size, which pushed down from the north covering New England, began to retreat from the Connecticut coastline. About 12,700 years ago they left Martha's Vineyard, an island off the southern coast of Massachusetts. Then these ice sheets uncovered what is now Boston about 12,300 years ago.

The peat samples led scientists to believe that 11,000 years ago the sea stood almost 200 feet below the present sea level. This left Georges Bank as an ice-free island about 100 feet above the water, covered with trees on the high land and salt marshes in the low areas. As the glaciers melted, the water covered the marshes and the low coastal areas. The part of the island that remained was carried away over the years by waves and winds and by strong tidal currents that are still common on the banks.

Fossil tooth dredged from the shelf has been identified as the tooth of the giant white shark, which is now extinct.
PHOTO: ROGER B. THEROUX, NATIONAL MARINE FISHERIES SERVICE

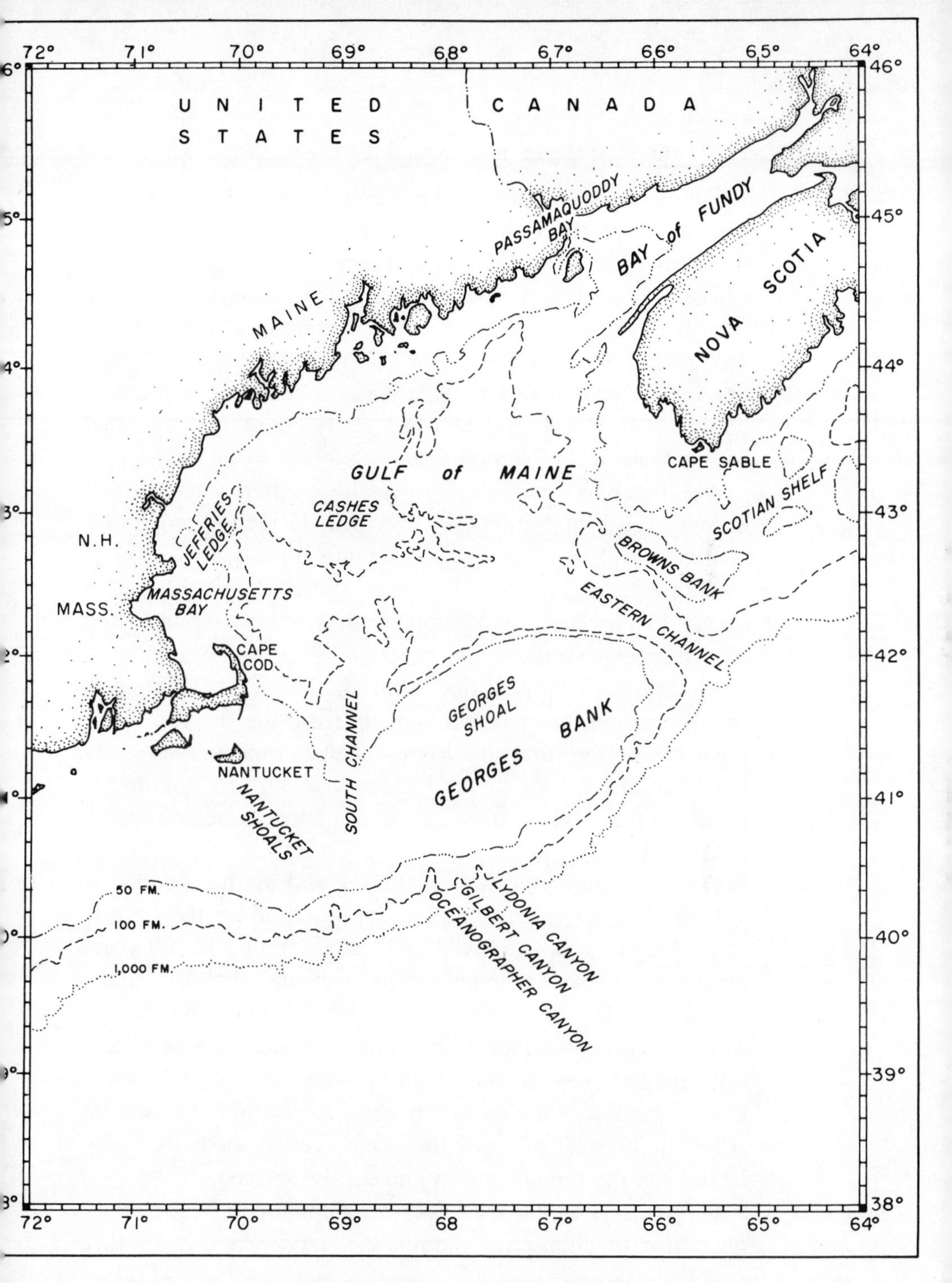

Location of Georges Bank, which was an island 11,000 years ago.
PHOTO: HERBERT A. ASHMORE, NATIONAL MARINE FISHERIES SERVICE

There is proof that Georges Bank was once an island, because a piece of fossil tree was snagged by the same scalloper *Ruth Lea* on the western end of Georges Bank near where the peat was found. The silicified limb of a tree (petrified wood) was identified as a section of a tree that was unknown to science. Two living relatives of the tree had been previously located in China, India and Japan. The piece of fossil tree from Georges Bank was only the second recorded of this kind in the New World.

A walrus tusk and mammoth tooth were found on the shelf in 1965, south of New York. The tusk, 15 inches long and 3 inches in width, came from an animal that lived 20,000 to 50,000 years ago. The mammoth tooth, almost 13 inches long, 2 to 5 inches in width and about 6 inches deep, weighed almost eight pounds.

In time it is hoped that additional objects concerning prehistoric man may be brought up from the shelves. Some scientists believe that the coastal shelves were inhabited by man during the time of the last ice age and are looking for signs of this. Items they hope to uncover include teeth, mollusk shells, animal bones and primitive tools that the inhabitants used for eating, gathering and hunting food.

People of the Clovis culture, characterized by their fluted stone projectile points, lived in North America 12,000 years ago. Since there was much game, fish and shellfish in the lowlands, it is reasonable to expect they lived on the shelf as well as in the woodlands. This thought was buoyed by the find made by a scientific party from the Woods Hole Oceanographic Institution aboard its research submarine *Alvin.* In 125 feet of water they came across shells that may have been the remains of an ancient oyster feed.

Of course not all shelf-floor samples are brought in by fishermen or submarine. Samples can be taken in another

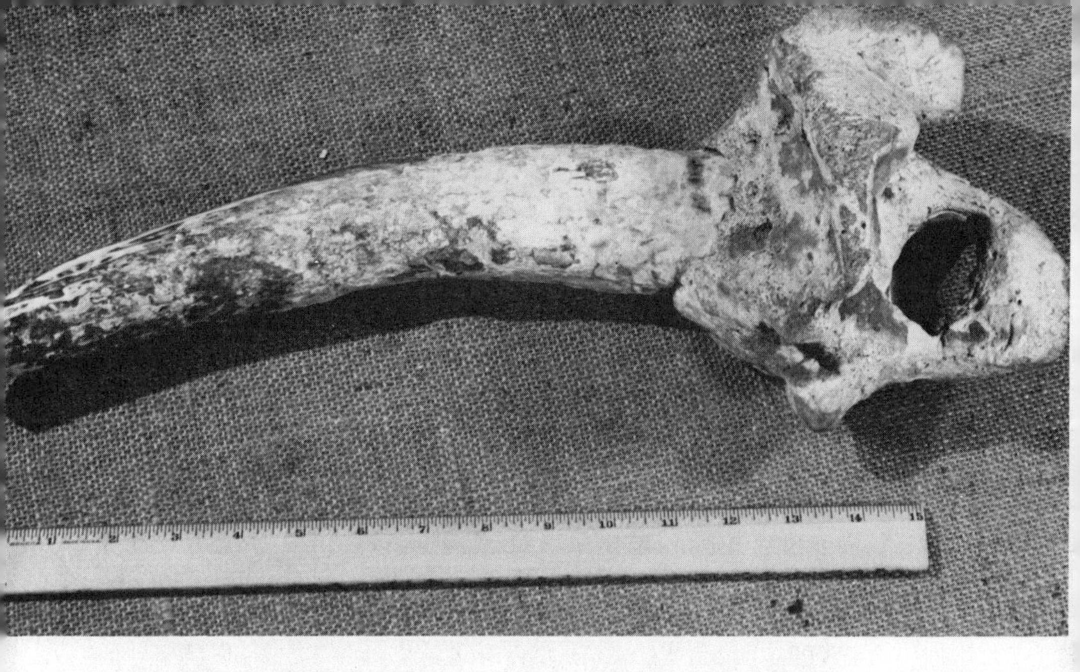

Fossil walrus tusk dredged from the shelf. Large part at right is a section of the skull.
PHOTO: ROGER B. THEROUX, NATIONAL MARINE FISHERIES SERVICE

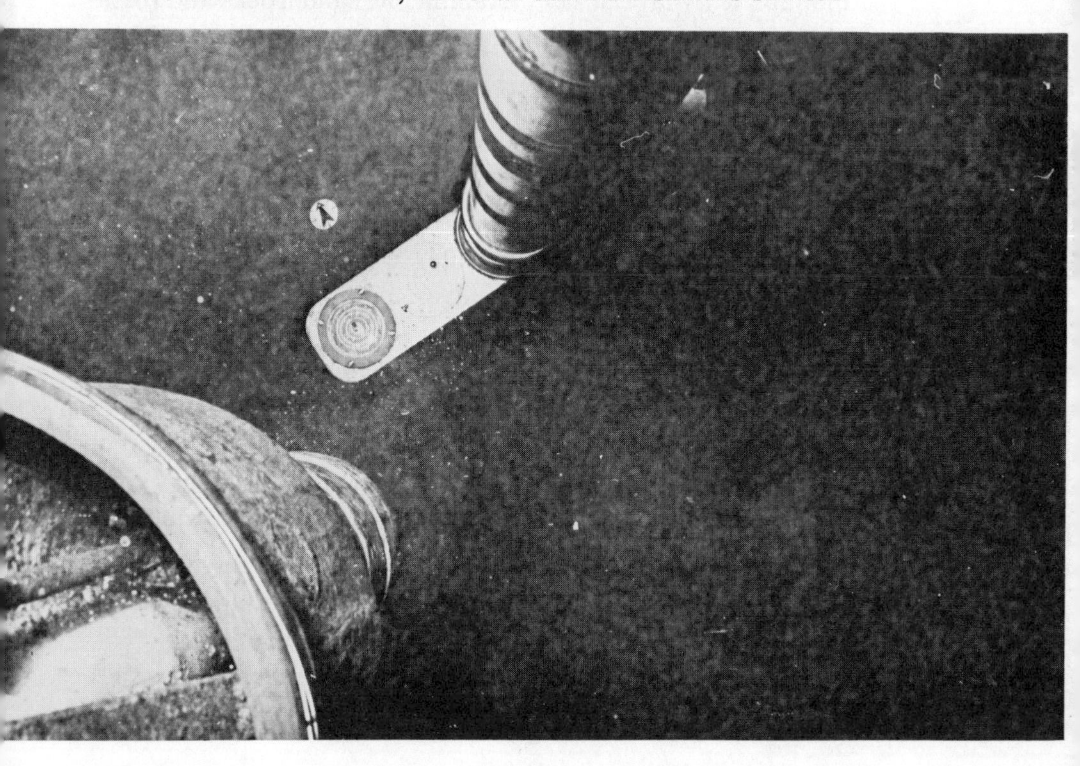

Underwater photograph of a corer, about to strike sediment on the shelf floor.
PHOTO: F. W. McCOY, HARVARD UNIVERSITY

way, a method called coring. A hollow pipe is actually driven through the sediment and 20 or 30 feet of the bottom can be encased, hauled up and examined. A heavy weight attached to the end of the pipe helps to drive the pipe into the ooze. Cables attached to the ship and the pipe allow the crew to keep control. When the sediment sample is inside the pipe, the whole coring rig is raised to the deck of the ship. The core is then pushed from the pipe, cut into small sections and immediately wrapped in air- and watertight wrappings to keep the core samples from drying out. The sample is then placed in a tube and marked as to where it was taken and when.

Once back on land, and in a laboratory, the core sample is removed from its container and examined. Geologists study the sample by layers. They examine the sand, rocks and fossil plants and animals in different layers of each sample. They measure the layers and compile the information. Some layers might have volcanic ash or bits of rock, suggesting a violent volcanic eruption that occurred sometime in history. The ashes drifted in the air, settled on the ocean water and then sank to the bottom, becoming part of the sediment. By studying these particles the geologist may be able to determine how the continental shelves in particular areas were formed.

Micropaleontologists, men and women who study tiny fossils found in the core samples, are a relatively new breed of scientists. The fossils they examine are very tiny, a few microns in diameter, with the average size one two-thousandths of an inch. They are called nanossils and include such organisms as rhabdoliths, coccoliths and discoasters.

In 1836 coccoliths were discovered by C. G. Ehrenberg as he was examining samples of chalk. At the time, he thought they were pieces of crystal and therefore not part of a living

A core sample being wrapped for later study.
PHOTO: WOODS HOLE OCEANOGRAPHIC INSTITUTION

organism. Later, in 1861, H. C. Sorby studied the chalk samples and noticed that the coccoliths were curved as though part of a small sphere. After a time, a complete sphere was found and it was covered with tiny platelets. This meant that the sphere was the skeleton of a living organism and not a crystal after all.

A bit later, in samples from the North Atlantic, G. C. Wallich and T. H. Huxley found additional skeletons, and Huxley named the small platelets coccoliths and the spheres coccospheres. Later, living coccospheres were collected as they swam on the ocean surface and were examined under a microscope. Then by 1900 it was decided that coccoliths were small platelets that made up the outside skeleton of calcium-containing plants called algae.

In all the oceans today, except in the cold waters near the poles, living coccolithophores live in great concentrations near the surface. They are a vital and early part of the food chain in the sea. They are one cell and are related to large seaweeds, but have characteristics of both plants and animals. They swim around freely and have a single whiplike arm, called a flagellum, that helps them to move through the water. Some live on the ocean surface and use sunlight in the process of photosynthesis just like larger plants. A few have been found in deep water indicating that they do not depend on the sun alone for existence.

In the 1950s scientists began to use fossil coccoliths in the science of biostratigraphy (which is simply the study of geographical layers or strata) by examining the fossils of plants and animals found in them. Most of the earth's surface is covered with sedimentary rock, and in the rock are small particles either living or nonliving. When a sequence of strata is studied, the history of the fossils can be recorded and used to determine the age of the rocks. Then, when the fossils of one sample of rocks are compared with those of another sample, it is possible to compare strata of the same age. With this knowledge scientists can determine what ancient climates were and have a good idea of what the environment was before man ever existed on earth. Other layers, depending on where they were gathered, may suggest other disturbances that took place long ago, such as ocean currents changing course, giant earthquakes or buildup of glaciers.

Since coccoliths are so very tiny, an ordinary light microscope, the kind found in most laboratories, is not adequate for examining them. Therefore little study was done on them until the development of the electron microscope, in 1932, and the use of electron micrographs some twenty years later. Using these instruments, plus the scanning electron microscope, it was learned that many of the coccoliths were alive 190 to 195 million years ago. Relatives of the coccoliths, called discoasters, became extinct 500,000 to three million years ago. With these new methods of studying fossils from the sea, many gaps of information may be filled in the geology of the shelves, the world and its ancient history.

There are also man-made objects in the shallow waters of the shelves that make interesting study. Most of those found on the continental shelves are in water shallow enough so that study and possible recovery is not unrealistic. There are such items as sunken ships, lost cargoes and parts of cities that dropped into the sea as a result of earthquakes. The rewards of this type of study, called marine archaeology, are great because certain periods of history can be revealed in more detail, explaining the type of ocean commerce, naval architecture and warfare plus forms of culture existing during those periods.

Archaeology, until about two decades ago, had been mostly a land-based operation. But the very equipment that increased the scientific exploration potential of the sea also opened up the field of ocean archaeology.

Problems facing the present-day underwater archaeologists are much the same as those being mulled over by land archaeologists. There are several things that have to be done. One is to use a regular system for exploring the find, whether it is a sunken ship or a lost city. Scientific drawings must be made. Each article must be identified and preserved. And, finally, the results of the expedition must be published. But, as could be expected, there is one further problem for the underwater explorer, that of reaching the site in the first place.

Before an archaeologist even thinks about how to study an underwater treasure, the site must be located. There is a great inner drive in some men and women to find sunken treasure and this urge often causes them to spend countless hours searching for something when they have no positive proof that it is actually there.

In the past, location was aided by rumor or a tale that was passed down from generation to generation. More reliable was the lucky fisherman who hooked onto a vase, anchor or piece of deck and hauled it up to his boat and then rushed back to tell of what he had found.

Today there is more sophisticated equipment. Metal detectors and small submarines help to locate underwater shipwrecks. Most wrecks are virtually covered with sediment or sand so there is little to see with the naked eye. One instrument that can be used in finding a possible site by analyzing the ocean floor is a hydrophone array that makes the task of profiling (making an outline of the ocean bottom)

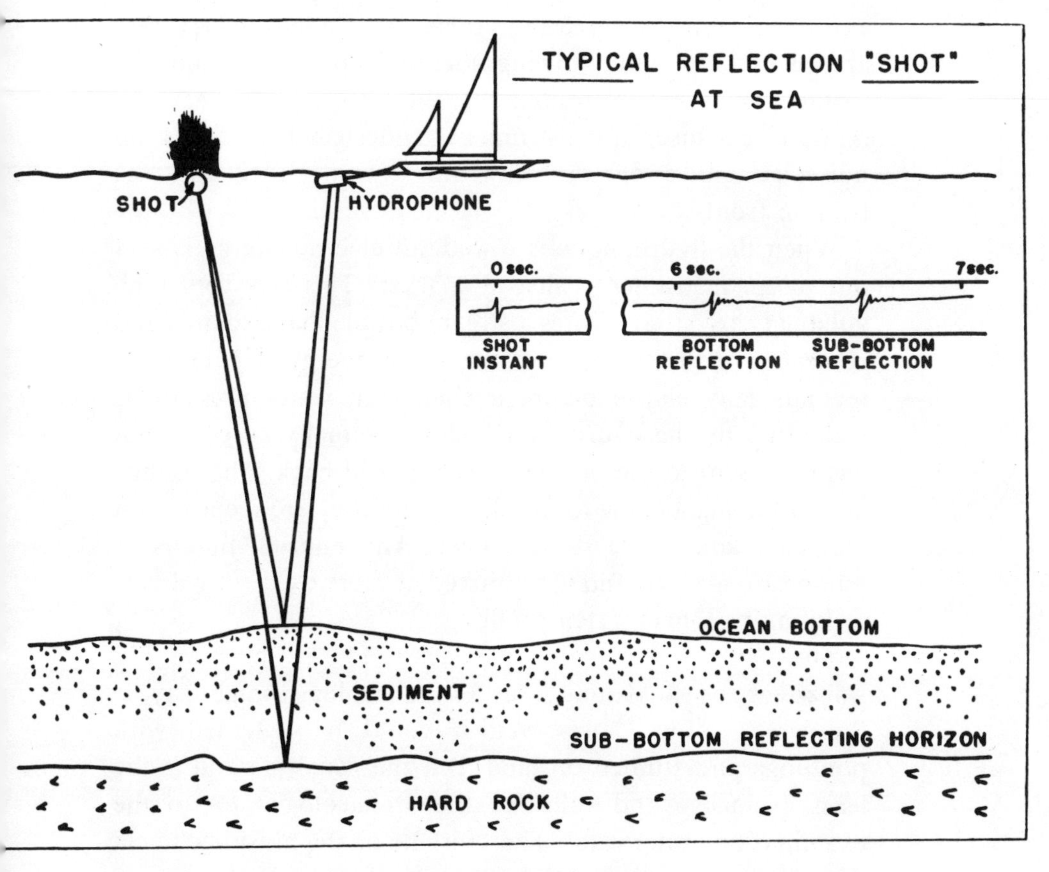

Explosive charge set at sea is measured by hydrophone picking up echoes as they glance off the ocean bottom and sub-bottom.
PHOTO: WOODS HOLE OCEANOGRAPHIC INSTITUTION

simpler and more productive than before. A plastic, called urethane, covers hundreds of tiny sensors, ten times more than used in previous hydrophones. This plastic, plus the additional sensors, greatly reduce the added noises that are picked up while towing the hydrophone behind the surface vessel. The sounds of the ship's propellers are reduced because the instrument is designed to focus on noises straight above or below and does not pick up noise from in front.

When the hydrophone is towed, an electric charge is sent out by an instrument called a sparker. This is a very high voltage charge that creates a stream bubble that expands and contracts in the water, creating more energy. Up to eight seconds may elapse before a change in water pressure is picked up by the hydrophone. This is actually an echo that has been sent to the bottom and bounced back. The sounds are sent along a cable to the ship where they are recorded on magnetic tape to be studied later. Any unusual bumps or humps are marked and then looked at more closely by divers or by more concentrated profiling.

There are a few main approaches to salvaging an old shipwreck, or excavating it, as an archaeologist might say. If the wreck is well preserved it could be refloated with pontoons and studied on land. But the usual method is the least expensive and calls for the archaeologist to do the examination underwater. Therefore he or she must become a diver, and a good one at that, since most wrecks are many fathoms deep.

Once a shipwreck is located, it is mapped or charted. This can be done by divers hovering over the wreck and marking what they see. Photographs are also taken with underwater cameras from above and from the sides, in an attempt to give an indication of how deep the shipwreck lies and what might be found at different layers. This, however, may take as long as two months when relying solely on divers to do the recording.

The University of Pennsylvania Museum, Underwater Archaeology Section, has a research submarine used exclusively for archaeological research. It is called *Asherah* and, using cameras mounted on its nose, it is able to photograph a shipwreck from above and from the sides in less than an hour, supplying information needed for a successful charting.

After mud and sediment are cleared away by divers, they use an airlift to raise the artifacts. This is a piece of flexible pipe that operates like a vacuum cleaner, sucking up small objects to the surface. Sand, mud and other unwanted material is washed away, while the larger and more solid objects are screened and kept separate. Large pieces, which remain underwater after the vacuum cleaning, are tagged. If they are small enough, they can be hoisted to the surface by way of an elevator, or basket with a rope tied to the handle. Larger objects are raised in an unusual fashion. A deflated plastic balloon is lowered to the object and tied to it. Then air is pumped into the balloon, causing it to rise slowly carrying the piece, perhaps a statue or cannon, to the surface.

Great care is taken to treat all objects properly once they are raised to the surface, as they may have been immersed in water for hundreds of years. Quick drying out could destroy them. Most objects are kept moist and allowed to dry slowly. Some are treated, especially wood items, by covering them with a substance called polyethylene glycol to help preserve them, since these items may have been attacked by worms. A soluble is gently forced into the empty cells to replace the water and keep the object from collapsing as it dries. Some metals must be treated, but gold is not affected by salt water and looks brand new when brought to the surface.

As objects are airlifted to the surface and more of the wreck is exposed the charting and mapping continues. The exact depth of each excavation is recorded and vertical measurements are made of all that remains.

Much has been found and charted on the shelves. Off the coast of Greece an ancient sunken city was located. A grain-loaded cargo ship that sank about the year 200 B.C. was also found there. Off the Turkish coast, thirty-five ancient wrecks were spotted and a Bronze Age ship of about the year 1200 B.C. was partially excavated. The remains of the *Bounty*, once skippered by Captain Bligh of *Mutiny on the Bounty* fame, were found off Pitcairn Island in the South Pacific.

Off the coast of Bermuda skin divers located gems, gold and historical artifacts that have been hailed as one of the most important finds of the Tudor Period, the fifteenth and sixteenth centuries. And tons of artifacts were raised from the sunken city of Port Royal, Jamaica, which fell into the sea during an earthquake in 1692, ending on the shelf floor. Many pieces of seventeenth-century pewter were added to collections.

One of the most remarkable feats of recovery in underwater archaeology took place in Sweden. In the year 1628 the Swedish warship *Vasa* sailed from Stockholm on its maiden voyage. A mile from shore it capsized in one hundred feet of water. Thirty-six years later, a Swede, Hans Albrekt von Treileben, raised fifty bronze cannons by going down in a diving bell that had air trapped inside.

After this recovery little else was done to the wreck until three hundred years later. Then Anders Franzen, an engineer, discovered that the Baltic Sea was so low in salt content that the pesky shipworm could not survive in that area. This meant that the sunken *Vasa* should have remained virtually intact. In 1961, after many hours of work and an expenditure of more than several million dollars, the ship was raised and is now on display in Stockholm.

The *Concepción*, a Spanish vessel that went down in 1641, has been located. It carried half the year's loot that the Spanish conquistadores took in Mexico. It is on a coral reef in the West Indies and has a value of precious metal, mostly gold, valued at anywhere from $50 million to $200 million. It is about sixty miles north of the Dominican Republic.

There is little doubt in the minds of women and men that the continental shelves have great historical value. The shelves offer a wide range of possibilities to keep inquisitive people busy for years in the way of discovery. And the tools of discovery are at hand.

3 How the Shelves Are Explored

No doubt when early man first walked beaches he was as curious then as we are today about what goes on underneath the endless waves. He probably took his first look at the shelf bottom through blurry eyes and, at the same time, realized that exploring underwater is much different and far more difficult than exploring on land.

Early man probably learned quickly, on the first dive, that he could not remain underwater for any length of time before he had to come up for a breath of air. And, if he dove down a few feet, he no doubt felt the weight of the water. Water weighs quite a bit, as anyone who has shoveled wet snow or carried a pail of water knows. This weight, or water pressure, presses on anyone or anything on all sides as it swims or crawls within the water. Even in the relatively shallow waters of the continental shelves the pressure against a swimmer is so great that he must use special instruments and devices to explore in safety.

What is true for the underwater archaeologist is true of the undersea explorer. The sea is darker, colder, more corrosive and slower to work in than the atmosphere on land. And it also costs the explorer more in time and money. Simple tools that men and women can use on land do not hold up to the tarnish, rust and dissolution that occur in sea water.

In fact, even before exploration of the shelves can take place with any kind of system and thoroughness, the explorers must know that they will return safely. They must also be reasonably sure they can find what they are looking for, and then be able to see what is there and record the necessary information once they do find it.

In shallow water free divers can use the self-contained diving apparatus. This consists of air tanks strapped to the back of the diver. Also used is a face mask for clearer views and rubber suits to help keep cold out and heat in.

Heat loss is an important factor in free diving in ocean water. Commercial divers have stated that they feel that they lose about 15 percent efficiency when they work underwater as compared to working on land. Scientists agree, having found in tests that cold promotes loss of memory, creates extraordinary dullness in the mind and impairs concentration. In a series of tests given to divers, three controlled situations were used. One test was on dry land, another in 64°F. water and a third in 45°F. water. It was learned in the experiment that the performance of a person in water was determined by two "effects," water effect and cold effect.

The water effect is the difference involved in movement in water, because of its density, compared to work on dry land. The diver is buoyant, unstable, senses are impaired because of the equipment worn and the resistance of the water slows all movements.

The cold effect is evident when all of the water effects are greater because of the colder water. The sense of touch leaves

Diver wearing self-contained diving apparatus, including air tanks and rubber suit.

PHOTO: UNIVERSITY OF PENNSYLVANIA MUSEUM—NATIONAL GEOGRAPHIC SOCIETY EXPEDITION

the fingertips and the strength of the hand grip is greatly diminished. Even the motion of the arms slows down. Memory is lost to a great extent and there is difficulty in doing tasks that are complex. Thinking is disrupted.

To try to help the diver in underwater work, heated suits have been developed. But they have their problems. If the diver needs to be attached to an outside heat source by a cord, then range and movement are restricted. Those suits that have their own self-contained heat units are often heavy and easily damaged. This hinders the diver's movements even more. The result is that it is difficult to explore the shelves by free-swimming divers.

Another way to explore the shelves is by echo sounding similar to methods used by archaeologists. A device that emits a clear sound is used. The sound is directed toward the bottom and the echo that returns is recorded. By knowing the speed of sound in water, instruments are able to measure the depth of the bottom and also the depth of the sediment deposits. What the scientists end up with is a profile that reveals canyons, hills, mountains, large rocks or any other objects along the shelf bottom.

Salt domes are found in this manner of profiling. One prime area where they exist is in shelves under the Gulf of Mexico. Here are the largest deposits of salt in the world, well worth finding and mining. The deposits occurred about 200 million years ago when the climate of the present site of the Gulf was hot and dry. It is believed that, at the time, ocean water flooded in easily but due to poor circulation did not flow back out. Because of a high rate of evaporation, and a rich salt flow from rivers, the water had a very high salt content.

Scientists also believe that during this time a volcanic dam or coral reef was in existence, causing the salt to build up into a thick layer. Then, about 130 million years ago, the floor of the Gulf began to sink and the barrier sank with it. From that time until the present, the salt layer has been slowly covered with a steady flow of sediment. Salt that was once just a few feet from the surface is now squeezed under 30,000 feet of sediment.

Pressure caused the formation of the salt domes. The thickness and the weight of the sediment, and the differences in the density of the sediment, caused the layer of salt to flow in one direction or another and eventually to grow upward. Sometimes earthquakes caused the salt to become a dome or at least be the start of one.

Once the salt began to rise, it cut its way through the layers of sediment until it was two or three thousand feet from the surface of the sediment. Here the density of the surroundings was less and the head of the dome began to flatten out and spread into the shape of a mushroom. This was a true dome shape, but others grew in many sizes and shapes resembling teardrops, ridges, pillows and cylinders. Major oil companies know of the more than 300 domes along the coast of the Gulf of Mexico because oil is usually located near them.

Profiling is not the only means of exploring the shelf. A more detailed means is by underwater camera. Cameras have been extensively developed in recent years but, compared to taking photographs on dry land, underwater photography is still a relatively primitive means of obtaining information.

Underwater cameras can be no further than 100 feet away from the subject to be photographed, and this distance is usually rare and only possible in the clearest of tropical waters. Most photographs are taken closer, meaning more photographs must be taken of a large area. Also, the results of underwater photography produce no sharp photographs. The limitations occur because light is absorbed by the water

An underwater camera, attached to weight in left corner, took this photograph.

PHOTO: U.S. DEPARTMENT OF THE INTERIOR, GEOLOGICAL SURVEY

and mixed with the tiny organisms that swim and drift along, causing it to change color or to scatter the light. To take a clear picture of a section of the shelf floor that would be only ten feet square requires that the camera be placed neither too far nor too close to the bottom in order to keep everything in focus.

Cameras are lowered by cable from a research ship and activated by using a sonar signal. The sonar indicates how far the camera is from the bottom. Two signals are sent from the camera. One is sent to the bottom and the echo returns to the ship. The second goes directly to the ship. By measuring the time difference of the two signals, computers can tell exactly how far off the bottom the camera is. The cable can be stopped when the camera reaches the desired distance.

No matter what the depth, water acts as a filter. As a camera is lowered into the ocean the thickening layer of sea water filters the entering sunlight. First the water absorbs the red, orange and yellow rays of light until, at deep depths, the warm colors have disappeared. At 50 feet red and orange appear as maroon or brown. A color picture at this depth taken with natural light would show only blue. To offset this, a filter is used that absorbs the blue and green, allowing the warm tones to come through. At very deep depths, filters will not work and artificial lighting is used. It is difficult to determine when to use flash. Of course, in the blackness of the very deep, powerful lights are a necessity. But, when the light of the sun is still visible, there is a difficulty in determining if artificial light is necessary. If too much light is used, the background becomes black.

Refraction, the changing of direction of light rays in water, makes everything look larger and closer than it really is. Therefore, an underwater photographer tries to use a lens that will correct the distortion that the water is making.

When a wide-angle lens is used underwater, it corrects the effect of the enlargement, providing a picture that is almost true. But, because the light rays bend the greatest around the edges of the picture, special lenses for underwater are used. These lenses correct the distortion and make underwater photographs fairly accurate in detail.

Newly developed cameras take 3,000 frames or more in a single lowering and are able to reach the lowest point of the sea fully corrected. Each frame records in a chamber the time, date, depth, film type and location.

For shallow diving photography, Jacques Cousteau, a French ocean explorer, developed a hand-held camera that can be used to a depth of about 150 feet. It is a 35 mm camera capable of shooting many frames. Often a 28 mm lens is used to insure more accurate results.

Movie cameras shoot 400 frames per second and up, and are good to depths of 1,200 feet or more. They hold 400 feet of film in one loading. Special lenses give the cameras a wide-angle field and the parts are nonfouling, shock resistant and powered with rechargeable batteries.

Use of stereophotography, in recent years, has improved the photographs taken on the shelf floor. Two cameras are assembled and placed with lights and other electronic gear that allow each camera to take photographs of the same area at the same time, in slightly different locations. The result is two overlapping photographs that tell anyone studying them the differences in height, depth and width of what is being photographed.

The two-photograph system can measure ripple marks in the sand as well as humps or mounds that may be the burrows of animals. Even the animals themselves can be measured along with calculations of mineral deposits.

Underwater television cameras are similar tools. If a

scientific party is interested in fish, and wants to observe them as they swim in their natural surroundings, they can lower a camera with high-powered lights attached. After the fish become accustomed to the camera, the scientist sits topside and watches what is taking place on a closed-circuit television set. Other television cameras can be fastened to tripods on the bottom so that divers can be seen from above or bottom equipment can be checked. Cameras are also used to spy on other animals and plants or to help locate objects.

Television cameras are also attached to the newest and most expensive instruments developed for underwater research, deep-water submersibles or submarines. These vehicles are capable of taking a crew of two, three and more into depths approaching two miles, and the stay is often many hours. Since the continental shelf is no more than 1,700 feet deep, the entire shelf is within the limit of exploration by most of these submersibles.

All but a few of the submersibles consist of a ball-shaped pressure hull crammed with instruments. Surrounding the hull is an outside skin, usually made of plastic or fiberglass material, that houses the outside controls such as propellers, mechanical arm, catch basket, television cameras, movie and still cameras and lights. The hull material is usually steel or glass, about an inch thick, able to withstand outside pressure of many thousands of pounds per square inch. Portholes of thick glass or plastic number from three on up, depending on the size of the vessel.

The crew of a three-man vehicle will consist of a pilot, copilot and a passenger, perhaps a scientist who is interested in the plants and animals or the type of rocks and sediment found on the shelf. The pilot uses two side propellers for steerage and stability. A large propeller in the rear provides forward movement. When the propeller is run in reverse, the submarine will move backward. The controls are simple in design so that incidence of failure is at a minimum. But they are very effective in a watery environment.

Motors on board are powered by batteries, and there is enough reserve power to run the vehicle three times longer than the prescribed length of a dive. There are oxygen bottles aboard and devices that filter out carbon dioxide within the pressure hull.

When a dive is started, the outside hull is filled with water and the whole vehicle becomes heavier than water. A descent starts. It takes only about fifteen minutes or less to reach the shallow floor of the shelf. Then outside ballast, a lead weight, is dropped to make the submersible neutrally buoyant. It will neither sink nor rise to the surface. The pilot then activates the propellers, levels the ship and proceeds with the prescribed dive. If the dive is deep or the water is murky, the outside floodlights will be turned on.

If a survey of the continental shelf is part of the mission, outside movie cameras record what is seen on the bottom. Samples of rock and sediment are possible with the aid of a mechanical arm. The arm has a scoop on the end and can dig into the sediment, take a scoop and place a container of sediment in an attached basket-type device on the outside of the hull.

If anything unusual is spotted on any dive, a photograph is taken and many feet of movie film are shot. Most of the film is color but those on board will also record what they see in notebooks so that their personal observations can be compared with what the films show, once back in the laboratory.

Diving in submersibles can be worked closely with fishery biologists. Submersibles dive to an area and scientists spot schools of fish by direct observation. While on the shelf floor they can also make direct ecological studies of the movement, distribution and habitat of the fish. The observers also study the bottom topography and composition to see how it relates to the fishing program. The fishing gear, including nets, can be observed firsthand to see how they function.

On-the-spot surveys of fish stocks are made to determine the populations of the schools and the possibilities of fishing in the area of the dive. The behavior of animals found is studied, as well as how they react to any large-scale disposal of waste products in the sea.

Submersibles are also used to map the continental shelves and to pinpoint the many ridges, mounds, canyons, shoals and depressions. Even though the shelves are mapped out by way of echo sounding, the returning echo cannot identify the type of sediment that is covering a mound or the types of animals and plants that may be living there.

In a typical dive the submersible *Aluminaut* was used to observe some royal red shrimp, in their natural habitat, to determine if they might be fished commercially. The descent was made in 250 fathoms (about 1,500 feet) of water into the shrimp beds directly under the Gulf Stream off Daytona Beach, Florida. Looking through portholes, observers discovered that the bottom was gray in color with sediment that was so loose it clouded the water with the slightest disturbance. They saw mounds and depressions with many holes in the sediment. Inside the holes they saw the claws of live animals, indicating that there were many crustaceans (marine animals with hard outer shells) living on the bottom. Sometimes fishes were seen in the holes. Very few of the fishes, crabs or shrimp were disturbed by the approach of the submarine.

The red shrimp were found at all depths ranging from 155 fathoms to 250 fathoms, with some concentrations as thick as

The research submersible Aluminaut.
PHOTO: REYNOLDS SUB/MARINE SERVICES CORPORATION

15 shrimp per 100 square feet of bottom. The average concentration, which was felt to be good for commercial fishing, was about 4 or 5 shrimp per 100 square feet.

Bottom photographs had shown previously that the red shrimp stayed on the sea bottom. Observation by submarine indicated that the shrimp were also in shallow trenches and that some were partially buried. Scientists believed the shrimp were buried because they search for food in the sediment.

The shelf is most thoroughly explored by living on it. Using the axiom that the best way to know a place is to live there, some oceanographers asked for and received a house in which to live on the floor of the shelf. At a cost of more than $2 million, an underwater house with four rooms was built and set on the bottom, at the edge of a coral reef near the island of St. John in the Virgin Islands. Thus in the original experiment, four men lived 50 feet below the surface of the sea for two months without seeing the sun or breathing fresh air.

The name of the house was *Tektite* and the project was done jointly by government and industry to see how well men could work and live for long periods underwater. As a result of the project it was determined that men could live and work under the sea for long periods. From that came other *Tektite* experiments, one with only women on board, and it was learned that women can live and work on the shelf as well as men.

While on the missions the experimenters did medical research, biological studies, made behavioral studies of each other and tried out various kinds of underwater hardware.

During the original two-month stay, one member studied lobsters and tracked them after he had tagged them with tiny sonar transmitters. He wanted to find out how they traveled and what their habits were in relation to day and night. Other members studied plankton and a coral reef.

Inside *Tektite* the breathing mixture had less oxygen than normal outside air and was also very humid, so the mixture would not support flame; this was for safety reasons. This meant that smoking was almost impossible, since a match would not stay lit and a pipe or cigarette could not be kept burning.

Mammals are being trained to assist aquanauts in their work as they live on the shelves. Bottle-nosed dolphins, harbor seals and sea lions are the new breed of aquanauts undergoing training to assist divers in search and rescue, delivering messages, mail, small tools, medicines, or to take samples topside.

In the role of rescue worker the animals are trained to swim to a signal produced by a diver who may be lost or in trouble. The mammal dives to the bottom, picks up an emergency breathing apparatus or hose and carries it to the alarm, thereby locating the person in distress. The mammal also carries a line to the diver who follows it back or takes a tow.

The signals are mechanically produced with a different signal for the type of assistance needed or the kind of emergency. Many buttons are attached to a diving belt, with one button meaning the diver is asking for more breathing mixture, another stating the diver is lost and a third requiring aid of another kind, such as the need of a tow.

Dolphins can tow many pounds through the water. In a test one animal was able to take 800 pounds, about three times its own weight, through the water. Heavy pieces of equipment could be used on the bottom, with sea mammals moving them along as the diver directed.

Locating lost divers could be the most important contribu-

Marine mammals, such as dolphins, are captured, studied and trained to help divers on the shelves.
PHOTO: MARINELAND OF THE PACIFIC

tion a dolphin can make toward the quest to stay underwater for longer periods and at greater depths. And dolphins could act as guards. One thought is to train dolphins to attack sharks on command. Another is to train them to attack other divers on signal if it is deemed necessary.

Perhaps it will be a matter of course to find that mammals of the sea will be working with man as the study of the shelf and sea continues. There is no doubt that the animals that live in the sea can do things that man cannot. A dolphin in one exercise was able to dive more than 200 feet, take a rope from one diver to another about 180 feet away and then return to the surface in only 70 seconds. It takes some bathers longer than that to get their feet wet!

4 Man and Shelf Animals

Whales, those huge mammals that are seen when they surface to blow out air and take in a new breath, are valuable not only as a friend in the sea, but in other ways as well. This is also true of the tiny barnacle that clings to the rocks and driftwood on shore. And such creatures as sea hares, starfish, octopuses and tiny dinoflagellates have a different value than one commonly thinks when their names are mentioned. Recently, new ways have been discovered to use these animals. They are being used in laboratory study, an important by-product of the sea, to increase man's knowledge of himself and his surroundings.

For years land plants and animals have been used in experimentation and development of drugs and medicines. There have been a substantial number of natural products gleaned from this land-based life, such as drugs to fight leukemia and heart disease, produce vitamins, help clot blood, cure common ailments and more. Today there have been additional discoveries concerning life in the sea on or above the continental shelves.

In two major ways the area of medicine has made practical use of the study of marine life. First, by using the relatively new field called marine biology, scientists study the physiology and structure of sea animals and plants and put this knowledge to use in the understanding of human disease. Second, with the discovery of the uses of chemicals that are produced by marine organisms, new medicines and drugs that act as tranquilizers, cure diseases or stimulate the human body, are among those being developed.

Land mammals, including man, have similarities but even remote animals found in the sea such as starfish or snails are valuable in the laboratory. Though human beings and sea animals are related only slightly, they are similar in the fact that they come from eggs. The egg is fertilized and then divides. By studying the simple sea animals, man can learn about fertilization and cell division. Much of what man does know today has come from the science called embryology, the study of the development of life in its first stages in the embryo. The research and experiments performed on such animals from the sea include worms, snails and sea urchins.

The squid, related to the octopus, is a popular animal in research and is often found in the marine biologist's laboratory. It has a long, thick nerve fiber in its mantle making it ideal for study. Probes can be inserted into the squid nerve and measurements taken as the nerve is stimulated. Inside each nerve is an abundance of material that can be squeezed out like toothpaste and studied for its chemical content. The human nerve fiber, on the other hand, is much too difficult to examine individually because the cells are so tiny.

The octopus has a well developed brain, allowing it to engage in certain kinds of learning. The whole nervous system of this sea animal is simple, and various types of

Researchers examine small animals netted off the shelf waters.
PHOTO: UNIVERSITY OF RHODE ISLAND

experiments, such as cutting off parts of the brain or treating sections of the brain with drugs, can be performed. But its brain is also complex enough that it is a simple and clear model for the human brain. In experiments biologists can show that certain responses, such as fear and rage, occur in the octopus. With this vast amount of study going on, it would not be surprising that some breakthroughs, due to experimentation with the octopus, might be possible in the treatment of human brain disorders.

Another sea animal studied is the sea hare, a small snail-like marine animal without a shell. It is also used in the study of the nervous system, as its system consists of only a few large nerve cells, compared to a human body, which has as many as 100 billion very tiny cells.

Marine mammals are also important in marine research. Whales have large, four-chambered hearts of such size that they are excellent for the study of heart diseases and heart malfunctions. Whale hearts also beat at a slower rate than the human heart, and this is of interest to heart specialists.

There is another important aspect in the study of marine mammals. Whales, dolphins, seals, sea lions and others are able to dive great distances under the sea without oxygen and can withstand the water pressure for long periods of time. Because man is also a mammal that wants to dive, marine biologists feel they should learn how and why an animal such as the Weddell seal, for example, is able to dive without apparent injury.

The Weddell seal lives in the coastal waters of the Antarctic continent and spends its time on ice and snow and in the cold waters above the shelves. It can live in air temperatures that drop to 70°F. below zero and swim in water where temperatures stay around 28°F. Ice covers the entire area for about eight months of the year, meaning the seal has to look

Dolphins are very important to marine research, and they are handled very carefully.
PHOTO: AQUATARIUM, ST. PETERSBURG BEACH, FLORIDA

for its food under the ice most of the time. Despite all these apparent disadvantages, the seal is able to grow fat.

The seal lives on fish that it finds under the ice. Being a mammal and breathing air, it must continually pop to the surface. It is able to find breathing holes in the ice and, if the hole freezes over, it uses its sharp canine-type teeth to chew away the ice. How it manages to find a single break in the ice in an area of many square miles is one of the mysteries to which marine scientists would like to know the answer.

Some marine physiologists, biologists who concentrate on the functions of the body, experimented with seals in Antarctica. These scientists wanted to learn how the seals found ice holes, how deep the seals dived, how long they stayed submerged and how fast they swam. More than one thousand dives above the continental shelves were observed, and it was discovered that the seals engage in three types of dives, each with a different purpose.

The first dive is brief, mainly for exploration. The seals dive for about five minutes to check the area and become familiar with it. If they run into other seals they usually have a fight to see who will be first to take a breath at the breathing hole. The second type of dive is deeper and longer, to a depth of 300 to 400 feet and lasting for about twenty minutes to an hour. During this time the seal can swim for up to five miles looking, perhaps, for other breathing holes. The scientists learned that sometimes a seal found another breathing hole and would not return to the previous hole. A third type of dive, one of very short duration, was to hunt fish for food. If fish were near, it did not take the seal long to find one and make a meal.

Scientists learned that seals and other marine mammals can stay in the depths for long periods and not need oxygen, due to the slowing of their heartbeat to a tenth of what it is normally. This reduces the rate of the metabolism and restricts the circulation of the blood to only the heart and the brain. In this way most parts of the body are not receiving any blood or oxygen; it is all being reserved for the vital organs.

The seal is able to withstand the tremendous water pressure it encounters in deep dives by not fighting the pressure but yielding to it. Its rib cage is very flexible and under pressure the lungs collapse, and the air is squeezed into the windpipe and into the cavity of the middle ear. In some sea lions, whales and seals it was learned that the middle ear is laced with many expandable veins, and it is thought that the veins swell with blood. When this occurs the volume of the middle ear is reduced, leaving no area in the seal's body with a great deal of gas. The result is that the body is able to avoid harmful differences between external and internal pressure.

It is possible for seals to become affected by the bends (when nitrogen gas bubbles are released into the bloodstream, cutting off oxygen to the tissues). However, seals are rarely affected because, when their lungs do collapse during a dive, there is little or no nitrogen in the lungs to be squeezed elsewhere.

It has been calculated that the lungs of seals collapse at a depth of about 160 feet, a shallow dive, and that seals should be safe in dives below that level. In shallower dives the Weddell seals usually come up for air at a very slow pace and they rarely make shallow dives of any duration. Therefore their diving habits plus their adapted anatomy keep them safe from diving dangers. By using this information plus much more, marine scientists hope to find methods to improve manned free dives to greater depths and in greater safety.

Public opinion has a bearing on what scientists study. The public is less interested in the behavior of a sea lion than it is in a cure or a help for a disease found in the pancreas of a marine animal. People feel that obtaining drugs from the shelf waters, by way of a science called marine pharmacology, is more dramatic than marine physiology. Therefore more work is done in the drug field.

It takes a long time to discover any drug, and it is equally as time-consuming finding something in the waters above the shelves that might make a contribution to mankind. In the past it has been popular to rave about the various medicinal qualities of the sea and the life found in it. For years people have gone bathing in salt water to heal their wounds, rid themselves of arthritis, clear their nasal passages, lose their dandruff or any other malady. In most cases it was too much to expect that any cure would be forthcoming.

On some occasions it has been proven that the old folk cures have some basis in fact. For centuries the Polynesians have eaten a particular kind of fish as part of a religious ceremony. After eating the fish, a person goes into a deep trance and awakens with the belief that he has died. Studies have been made and the results stated that a drug in the fish is a powerful psychedelic with various effects. It sends the user into a coma, causes depression and a suspension of the will. Little else is known of the drug but there are further investigations to see if it might be used in some beneficial way. By changing the dosage, a drug can sometimes be controlled and will prove useful. For example, when certain parts of marine animals are taken internally they can be poisonous. But if the poison is diluted, then what was once harmful can be, and often is, helpful. For many years Japanese doctors have been treating cases of poisoning in people due to their eating of *fugu*, a dish prepared from the meat of the puffer fish. Tetradotoxin, which is obtained from the puffer fish family, is one of the most potent poisons. As a result of extensive laboratory tests, the drug tetradotoxin is now used in dilution to help relieve pain in terminal cancer cases.

Another poison, cephalotoxin, taken from the saliva of the octopus, is a potent drug that, when used in diluted solutions, helps to control high blood pressure in man and helps regulate the irregular heartbeat in heart-attack victims. The poison from a shell animal called the cone shell is used to relax muscles and it is hoped that it might be used to stop convulsions. Another poison taken from the cone-shell family helps contract muscles that have been damaged by injury or illness.

Essentially the drugs that are being taken from sea animals for study and eventual use are divided into two classes, systemic drugs, which work on parts of the body, and antibiotics, which kill or destroy disease organisms. The systemics, although poisons in improper dosages, may act as antidotes, ease pain, speed up blood clotting, help to relax or stimulate muscles and promote healing. One interesting point is that most sea animals have a poison that aids them in killing or stinging an enemy or assists them in the quest for food.

This killing in the sea by way of secreting a poisonous liquid is called antibiosis and sometimes is quite spectacular. The famous red tide, where ocean water is turned red in color, is the result of tiny organisms, part of the plankton, that are called dinoflagellates. When conditions are right, and there is an abundance of food plus necessary light and temperature, the dinoflagellates multiply rapidly and dye miles and miles of ocean red. They live by the billions and leave behind a poison that is thought to be responsible for killing great numbers of fish and for causing shellfish in the area to be unsafe for humans to eat.

One poison found in dinoflagellates has been particularly effective, as an antibiotic, against many different disease bacteria. It has also been able to reduce blood pressure and is thought to be an excellent source for a drug that can be used to reduce hypertension.

Other drugs have been located in peculiar fashion. During a study in the Antarctic, scientists discovered that the intestines of penguins had little, if any, bacteria. By tracking this back to what the penguins ate, the scientists learned that a tiny shrimplike animal called krill fed on a particular kind of one-celled sea plant. This plant, one of the marine algae, produced a substance that was isolated in a laboratory and is called halosphaerin. It has been used to destroy staphylococci, potent bacteria affecting man.

The internal organs of the sea cucumber were found to contain a drug that was named holothurin. In tests it stopped the growth of some tiny protozoans and had effects on the development of the fruit fly. It also deadened nerves and stopped the growth of tumors in mice, but it did not have any antibiotic strength.

The sponge, another kind of marine animal, is an additional source of antibiotics. An effective agent called ectyonin was extracted from the red sponge. It would kill certain types of disease germs, one being a staphylococcus that resisted penicillin. When the extract from a green sponge was mixed with holothurin, the antibiotic power of the sponge was no longer effective. It is not known whether this information could be used for medicinal advantage or not.

The hagfish is an eel-like fish that has twelve pairs of gills, two rows of hard, yellow teeth and three hearts that are separated from each other. One of the hearts, in fact, is not even connected to the nervous system, yet beats in time with the other two. Study of the hagfish resulted in a drug called eptatretin. When the drug was injected into dogs in laboratory experiments, it caused damaged heart nerve centers to begin to regulate heartbeats again. It is thought that this drug may be used in regulating the heartbeat of patients who have heart disease or heart malfunctions. Further studies have revealed that this drug is only found in the hagfish.

There is a chemical that is taken from the venom of the weeverfish that can reduce the heartbeat of an animal to a few throbs a minute. This could be most helpful in surgery because, during operations, it is often necessary to tie off arteries to minimize blood flow.

A chemical taken from the mouth of a marine worm has the ability to stop the growth of sea urchins' eggs when the chemical is placed on them. It was added to cancer cells in a tissue culture and the chemical stopped the cancer growth. Perhaps it may be an important inhibiting agent in the fight against cancer.

Various fishes are good research tools because, even though they are an unlikely comparison, they do resemble man in certain ways. Fish contract tuberculosis and cancer; they are attacked by microscopic bacteria, protozoa and fungi; and they suffer from cirrhosis of the liver, vitamin deficiencies, and heart defects due mainly to fat in the tissues.

Transplants are taking place in fish. The molly is a fish that reproduces its kind without benefit of a sexual partner, and transplants of its heart, kidneys, ovaries, pituitary glands and other parts of the fish have been successful. There is hope that knowledge gleaned from this type of research may result in more successful transplants in man.

Certain sea animals are also being studied for various uses, concentrating on something they do in their natural life and

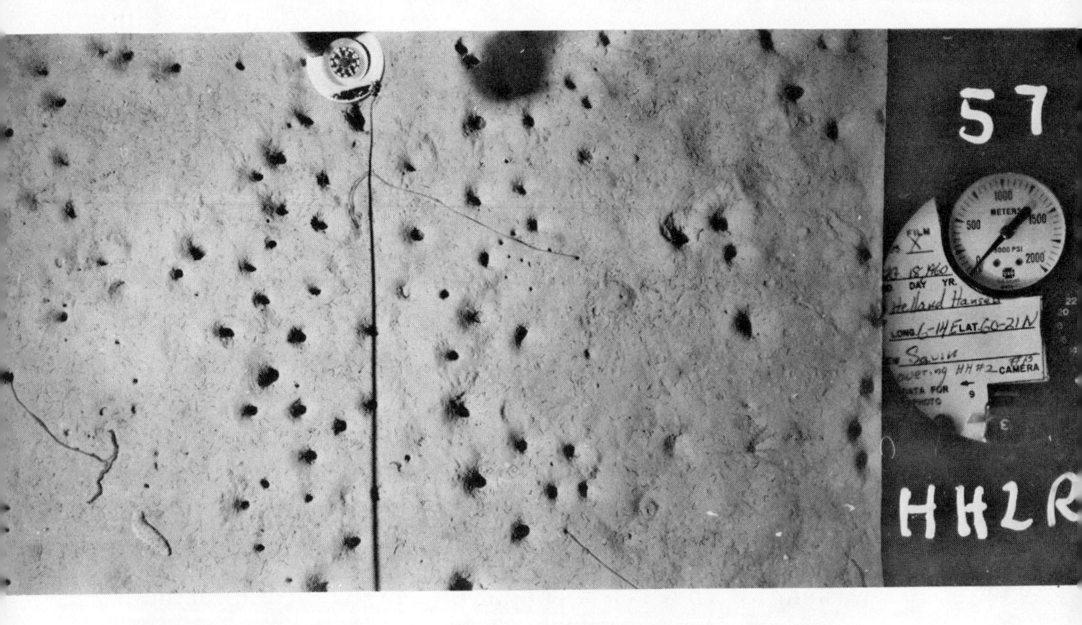

The long, thin lines are marine worms, photographed off the coast of Norway. Block on right gives date, location, camera, lowering number and type of film.

PHOTO: WOODS HOLE OCEANOGRAPHIC INSTITUTION

utilizing it for man. One such study is on barnacle glue. For as long as man has been sailing boats across the sea or has driven piles into the mud and sand, these tiny, white, hard-crusted and difficult-to-remove sea animals have plagued the mariner. Anyone who has tried to scrape them off the bottom of a boat can truly attest that they have a strong glue, because usually the boat paint is removed before the barnacle is sent on its way. It was thought that any adhesive that can hold an animal to a ship's hull, a piling or a rock, despite being constantly pounded by the surf of the sea, must have an unusually strong holding power.

Dentists maintain that there is no glue manufactured by man that can hold a filling in place. Therefore cavities must be drilled out larger than is really necessary with a wide base and narrow opening so that the fillings are locked in place. If there were a glue that could hold a filling, despite a constant washing of coffee, tea, soup, milk, the crunching of hard candy, the chewing of roast beef, and the pulling of taffy, then the business of dentistry would be much better off and so would the patients.

In experiments barnacle cement was put through a series of tests. In one test, for example, a layer of barnacle glue, thinner than a coat of paint, had a strength of over 7,000 pounds per square inch. This may not mean much until it is compared with the best epoxy glues that are used on spacecraft. The barnacle glue was twice as strong as the epoxy and four times as strong as the very best wood glue that is used around the carpenter shop.

There are other observations of the strength of barnacle glue. When barnacles were removed from steel, bits of steel were stuck to the base of the shell. There have been fossil remains of barnacles that attached to shells some 15 million to 20 million years ago. The barnacles were still attached after all that time!

In other tests barnacle cement was heated to 662°F. and did not melt. It was frozen at 383°F. below zero and the glue did not peel or crack. It cannot be dissolved in most strong acids, or any organic solvents or alkalis. The fact that it does not dissolve in water interests dentists and doctors because the cement actually cures in seawater, and the human body is about 70 percent saltwater.

Three or four times a year the adult barnacle spawns and produces about 10,000 to 15,000 eggs that are kept inside the barnacle shell, for about ten days, until they hatch. The larvae then drift around on ocean currents for a month feeding on plankton. As they drift they are also eaten. Those that survive go through seven stages and then hook onto something, usually somewhere between high and low tide where there is much sunlight and strong currents. They land head first and "walk" around on their antennae, making cement as they move. Soon they are stuck fast and spend their life on an old board, piling or rock.

The difficulties in obtaining barnacle glue lie in the small size of the animal. Thousands would have to be cleaned of their glue supply to get a few drops. Also, because of their size, they would be difficult to gather in great numbers. But, in time, scientists hope to be able to analyze the cement and then produce it synthetically.

There are some larger animals in the sea that are not used by man for purely scientific purposes. These are animals that are also caught or captured, kept alive, fed, trained and put in aquariums throughout the world for people to see and admire. These are animals such as seals, sea lions, porpoises, small whales, eels, sharks, turtles, snails, sponges and many kinds of small fish plus a variety of other animals and plants.

Animals by the hundreds are captured alive in waters above the shelves by people who spend hours or even days tracking down a particular specimen. They not only go after the usual or common variety, but they search out the unusual as well. In the 1960s it was learned that there was an all-white animal swimming about in the shelf waters of St. Helena, off the coast of South Carolina. It turned out to be an albino bottle-nosed dolphin.

It took three expeditions by the collecting crew from the Miami Seaquarium to finally capture the white beauty. In the first attempt she was snared in a net but tore through it and escaped. The second attempt failed to locate her and winter storms made it necessary to turn back and hope to return in warm weather.

Months later, on the third try, the collecting crew caught sight of her and made chase, which lasted for sixteen days. Then she was finally netted and brought aboard the collecting boat. She was eight feet long and weighed about 400 pounds. The flesh around her eyes was pink, as was her mouth. She was named Carolina Snowball.

The dolphin was taken back to Miami and placed in a $100,000 pool at the Miami Seaquarium, complete with picture window. The pool was built especially for her. She was considered to be the single most expensive aquarium specimen in the entire world.

She was viewed by millions, and scientists came from around the world to study this strange quirk of nature during the almost three years she lived in captivity. She died of many causes, none attributed to her life in the aquarium.

There are many other animals taken from the waters above the shelf and from the floor of the shelf as well. The value of the products taken from the animals, and the animals themselves, is in the many millions of dollars. What cannot be measured is the value these products have to many in everyday life in the way of life-saving drugs.

5 Fishing on the Shelves

Go out on a boat someday and maybe the old man of the sea will be there. He, with his seaweed hair, barnacle-encrusted clothes, seashell buttons, and riding a porpoise, may beckon to you to come over and see what is upsetting him. He may point a bony finger down into the water to a place called Georges Bank and shake his head. Sadness will crease his face and, if you look closely, you will see tears, salty tears, in his eyes. He will be crying because marine life on the continental shelves is changing. Where once fish roamed the shelf floor by the millions, only a few swim there today. The old man of the sea will stroke his whiskers and ask why? Can any of us give him a logical answer?

Marine scientists say the scarcity of fish is due to overfish, their terminology for taking too much of a species and not allowing time for them to breed in normal fashion to replenish their stocks. In the North Atlantic, overfish became very real—in fact it was labeled a crisis. And there are many stories about overfishing that are alarming.

As little as a hundred years ago it was felt by most people, including many scientists, that the fish in the waters over the continental shelves would provide an endless supply of food for the world's population. Fishermen could cast their nets, make their hauls and bring back necessary protein to their home ports. But, because of several factors, that one-hundred-year-old belief has proved to be very foolish. Many of the world's most vulnerable and valuable species of fishes have been fished at such a rate that they are declining in population.

Beginning in the late 1950s, for example, off the coast of Peru, fishermen brought home millions of tons of anchovies that were ground into fish meal. This meal was sold as high-protein feed for poultry and livestock. The fish were found in the Humboldt or Peru Current, a cold-water stream that abounded with food for the anchovies. Every seven years, on the average, a change in winds brought a warming current that flowed over the cold water, causing the fish to disappear for a few months. Peruvian fishermen knew this was temporary and didn't mind the delay.

In 1971 things changed. The warming current did not disappear in a few months. It stayed on and on and lasted for more than a year. The catch was very, very low. The government banned most fishing during 1972 but, when the boats set out again in 1973, they came back with empty holds. The anchovies had almost disappeared.

Fishery biologists explained that the loss of fish was not entirely due to the warm current. The blame was placed on the greed of the fishermen. Biologists said that a yearly haul of about 10 million tons would keep the supply of fish stable; the fish could reproduce themselves. But in 1971 some 11 million tons were landed compared to a record 12 million in 1970. Biologists warned that the stocks of fish were so depleted that it would take years for them to regain their size, if they could at all.

However, in 1974, the anchovies began returning to their grounds as mysteriously as they disappeared. Scientists are not sure just what is bringing the fish back. And they don't know how many are back and for how long. One thing they know is that the government has cut down the amount of fish that can be caught. And, with luck, the anchovies will be saved and can be fished in moderation for years to come.

Off the coast of New England, the northeastern section of the United States, the fishermen need more than luck. The sea herring, a onetime valuable food fish, is depleted to only 10 percent of its former numbers. Haddock, a most popular food fish, is approximately 3 percent of its normal population and some biologists fear it will never recover. In California striped bass have been reduced by one-half.

Why are these particular fish disappearing? One reason is that power companies, struggling to keep up with energy demands, construct complexes along tidal estuaries where fish, especially young fish, live. These power plants suck up larvae and eggs and, in so doing, are responsible for eliminating billions of potential adult fish. Another reason is that huge water-diversion plants throughout the world pull fish into irrigation canals and pump them to inland waters where fish have no chance for reproducing. Most important is that massive fleets of fishing vessels comb the floors of the shelves with their nets and scoop up fish by the billions each year. These fishing fleets are a story in themselves.

For many years the United States, Japan and Norway were the leaders in fishing, with Russia usually holding down fourth place. But around the early 1950s the Russians fully realized that they, along with the rest of the world, were approaching a food crisis. They felt that the sea would most

An American fishing trawler.
PHOTO: ROBERT K. BRIGHAM, NATIONAL MARINE FISHERIES SERVICE

likely help to alleviate the problem, at least for the time being.

At the same time the Russians were realizing the importance of food from the sea, the Japanese were rebuilding their nation's deep-sea fishing fleets. These had been destroyed or allowed to deteriorate because of their participation in World War II.

After the buildup, the Japanese fleets began bringing in more fish than they had brought in during the years before the war. They had new factory or mother ships that were supported by fifty or sixty trawlers, measuring about 150 or 250 feet in length. There were also tankers, cargo and supply ships. Research vessels cruised all over the Pacific Ocean hunting for new fishing grounds.

During this time the Russians still fished their prewar fishing grounds, often side by side with the Japanese. Occasionally a new factory ship or trawler appeared on the grounds, but this did not foretell what was to come. What did lead some scientists to wonder, however, were the constant reports of Russian survey and research craft appearing throughout the world in many places where Russian ships had never been seen before.

In 1954 Russia caught seafood at the rate of about 2.5 million metric tons per year, which was the same as she was bringing in four years earlier. Meanwhile Japan had soared to a high of 4.5 million metric tons. But then, in 1955, the massive changes came about. A large Soviet research vessel appeared in the northwest Atlantic to survey the rich fishing grounds that extended along the shelves from Greenland south to the shores of New York. In this area are the famous Grand Banks, located off the coast of Newfoundland, Canada, and Georges Bank, which is off the coast of New England. Both locations had long since been recognized by local fishermen as being two of the best fishing grounds in the entire world.

Japan, the United States, Iceland, Denmark, Canada and Great Britain usually fished these grounds and they all were surprised to see the Russian ships. But no one fully realized that this was the sign that signaled the Russians would soon be appearing in numbers.

Later it was learned that, at the same time, an entire new fisheries division was being established by the Russians in a port located on the Black Sea. This port would be responsible for fishing the fruitful areas in the southeastern Atlantic, the Indian Ocean, the Gulf of Aden, the Arabian Sea, the east and southwestern African coasts. This again surprised the world's fishing nations and foretold that Russia would join the giant fishing industry on the high seas, above the continental shelves, wherever there was an abundance of food fish.

Japan was most worried. She realized that she would be competing with Russian ships for the same fish in the same seas. And there would also be new competition for tuna, a fish that Japan has specialized in and had canned and exported to the United States in huge quantities. For years tuna brought in the highest prices. Russian ships were seen in tuna areas such as those off New England, in the southeast Atlantic and coastal regions along Central America.

Japan did not take long to act. She expanded her fishing fleets and spread out in the Pacific Ocean south and east of Hawaii and to the west coast of the United States in search of more tuna. During the expansion the Japanese developed the long-line method of catching tuna, utilizing lines that were up to ten miles long with hooks located every 200 feet. Japan, because of its relatively small land area compared to its population, had to turn to the sea for its food.

To get an idea of how a vast fishing fleet operates we'll examine the Russian fleet because it is most interesting and the most modern of any in the world. The Soviet fishing fleet has more than 20,000 large trawlers, factory ships, mother ships and refrigerator vessels. In their holds is a capacity to store approximately 3.5 million cubic yards of food from the sea. Most of this space, 2.7 million cubic yards, is used to hold and transport frozen fish and seafood. This massive storage area is needed because, in the space of twenty-four hours, the Soviets have a capability of freezing 18,000 tons of fish.

Some of the larger trawlers have huge 2,000-horsepower engines and the ships displace about 3,600 tons. They are complete with a stern trawl rig and refrigeration system. Each ship can hold 650 tons of fish in its bowels. During an average year the take for each ship averages about 10,000 tons of fish.

The fleet of large refrigeration ships has the capability to freeze 100 tons of raw fish, process 100 tons of scraps into fish meal and salt 100 tons of herring every twenty-four hours. These ships can travel about 10,000 miles without refueling and cruise at speeds of 20 knots (approximately 23 miles, or 37 kilometers, an hour). The temperature in their holds is kept at well below zero.

The canning ships have stern trawls and holds where the cooked fish can cool. In twenty-four hours all ships are capable of canning 200,000 tins of fish and turning raw fish into fish meal and fish oil. Each mother ship is able to process 300 tons of raw fish every twenty-four hours. When fully outfitted, each vessel can stay at sea for three months.

Whereas the Russian ships spend months over the shelves fishing, boats from the United States go out for only a few days at a time. The reason is simple enough; the American

Soviet factory stern trawler, part of their fishing fleet.
PHOTO: NATIONAL MARINE FISHERIES SERVICE

boats are closer to home and are much smaller. But the goal is the same and the method is similar too—tow the nets across the bottom to net the bottom fish.

There are different kinds of nets used by the American fishing boats; one that is popular with most nations is the otter trawl. It has otter boards or "doors" placed on either side of and forward of the net opening. The doors keep the net mouth open and improve the efficiency of the trawling procedure.

The net is towed by cable across the shelf floor, at depths of 180 to 240 feet, to catch ground fish that feed off the bottom. The ship glides forward and, with the help of currents, the fish are forced toward the rear end of the net, which is called the cod end. Once the fish are in the cod end, there is little chance they can escape. As the doors keep the net open, floats keep the top of the net up and heavy chains hold the bottom down.

If the net is torn while being dragged over a rocky bottom it is repaired as soon as it is hauled topside. Some fish are cleaned as soon as they are brought aboard, stored in the hold and iced down to keep them cold. Most of the fishing draggers work night and day, especially if they find the fish. No one wants to stop to sleep for fear they might lose the school.

Once home the fish are readied to be unloaded at the dock. If the boat is from the United States, the price of the catch has to be established before unloading. This is done at a fish auction, usually at a building close to or right on the dock. After the price is arranged, the fish are unloaded. A man goes into the hold and, using a short-handled pitchfork, loads the fish into wire baskets. He is called a fish lumper. He is paid so much an hour for unloading, and is experienced and fast.

The wire basket is hoisted by pulley to the dock and

Unloading an American fishing trawler. The wire basket has been loaded in the hold by a fish lumper.

PHOTO: ROBERT K. BRIGHAM, NATIONAL MARINE FISHERIES SERVICE

dumped while the fish lumper is loading a second and third basket. The fish are then sorted, according to size, and prepared for market. Some are iced in wooden boxes and taken to a wholesale market, where they are cleaned and filleted. Others are shipped directly to markets for sale to local stores where consumers make their purchases. This style of fishing and unloading is much different from that of the Russians and their factory ships.

Now that the reasons and methods of fishing on the shelves are known, what might happen in the future in the waters over the shelves? And how will it affect the old man of the sea? Will his unhappiness ease and will there be a reason for him to smile again?

In the fall of 1973, fourteen nations agreed to stop overfishing on Georges Bank, cutting back their annual catch by 25 percent. Also, fishing vessels longer than 150 feet will be prohibited from fishing in waters of less than 40 fathoms (240 feet) between July 31 and December 31 of each year. This will protect the New England fisherman by keeping large boats from local waters.

Now that something has been done to help protect Georges Bank, what about other areas of the world? A few years back Peru claimed as territorial waters 200 miles offshore. This takes in the Humboldt Current so, in essence, the anchovies were claimed by the Peruvians. When foreign fishing vessels are caught within the 200-mile limit by Peruvian patrol boats, they must be brought to port and pay a fine. All this is an attempt to help save the anchovy fishing for the Peruvian fishermen.

Iceland extended her territorial waters to protect her cod fisheries. England took exception to this and challenged the right of Iceland to make a claim. The result was a "cod war" causing seagoing animosity between the two countries.

Rammings were reported on the high seas as well as shootings and numerous threats. In time a compromise was reached.

New England fishermen are not satisfied with the agreement concerning Georges Bank, however. They want to follow in the footsteps of Peru and extend territorial waters of the United States to 200 miles. When and if this occurs, it will eliminate any foreign vessels along the entire eastern coast and, if Canada follows the same line, there will be no foreign fishing off the coast of Nova Scotia as well.

In fact some lawmakers and biologists want to close down the fishing grounds entirely for a few years. They offer the following as proof that their plan will bring back the fish. During the last two world wars there was a sharp drop in fishing because of necessary restrictions. Immediately following each war, when the fishing boats went to sea again in large numbers, there was a remarkable increase in catches compared to prewar averages. These experiences, and other similar examples elsewhere, point out that fish stocks can be depleted but they can also be reversed if the fishing stops or is cut back to allow the fish to replenish themselves. It is felt that, with proper management, stocks of fish could be built up to where they will be able to supply a good amount of protein to the world.

The danger of overfish can also bring trouble and despair to those who earn their living fishing and to the villages where they live. For example, one only has to look at the fishing industry of the state of Massachusetts, an industry that has been in business since the landing of the Pilgrims. The Massachusetts fisherman found he was being undersold in his own home ports by foreign imports and that he was being outfished by bigger and better fishing boats and more of them.

Some towns along the coast had their local fishermen turn to other types of fishing when areas ran dry. Fishermen turned to lobstering, shellfishing and long-lining.

Long-lining for swordfish is one alternative and the method is unique. Swordfish are found off both coasts of the United States and in most of the temperate and tropical waters of the Mediterranean and Atlantic. They are also found near Japan, New Zealand and Hawaii. They feed on butterfish, mackerel, hake, herring, menhaden, pollack, lanternfish, redfish, haddock and squid. They travel in schools but are usually no closer to each other than 40 feet. The key to the swordfishing industry is the temperature of the water. The fish prefer warm water of about 60°F. and higher. There is not much known about their movements but marine scientists believe that the fish migrate to deep water during the cold months and can dive to depths of 1,200 feet.

Fishery biologists think that swordfish spawn during the months between June and September, with mature females producing 10 million to 100 million eggs. They hatch in about two or three days and what happens after this is not known. But the average fish caught weighs between 100 and 400 pounds and some large fish have weighed up to 1,000 pounds.

Fishing for swordfish occurs toward the far reaches of the continental shelf where it begins to drop away and forms the continental slope. The depths range from 600 feet to 3,000 feet in sharp drops. The lines are baited with one-pound mackerel on hooks that are set about 10 feet or more apart. Several miles of line can be used with buoys and markers with radar reflectors every half mile to warn approaching traffic. The important fact about swordfishing is that it is a popular fish and usually brings a good price at the market, regardless of the time of year.

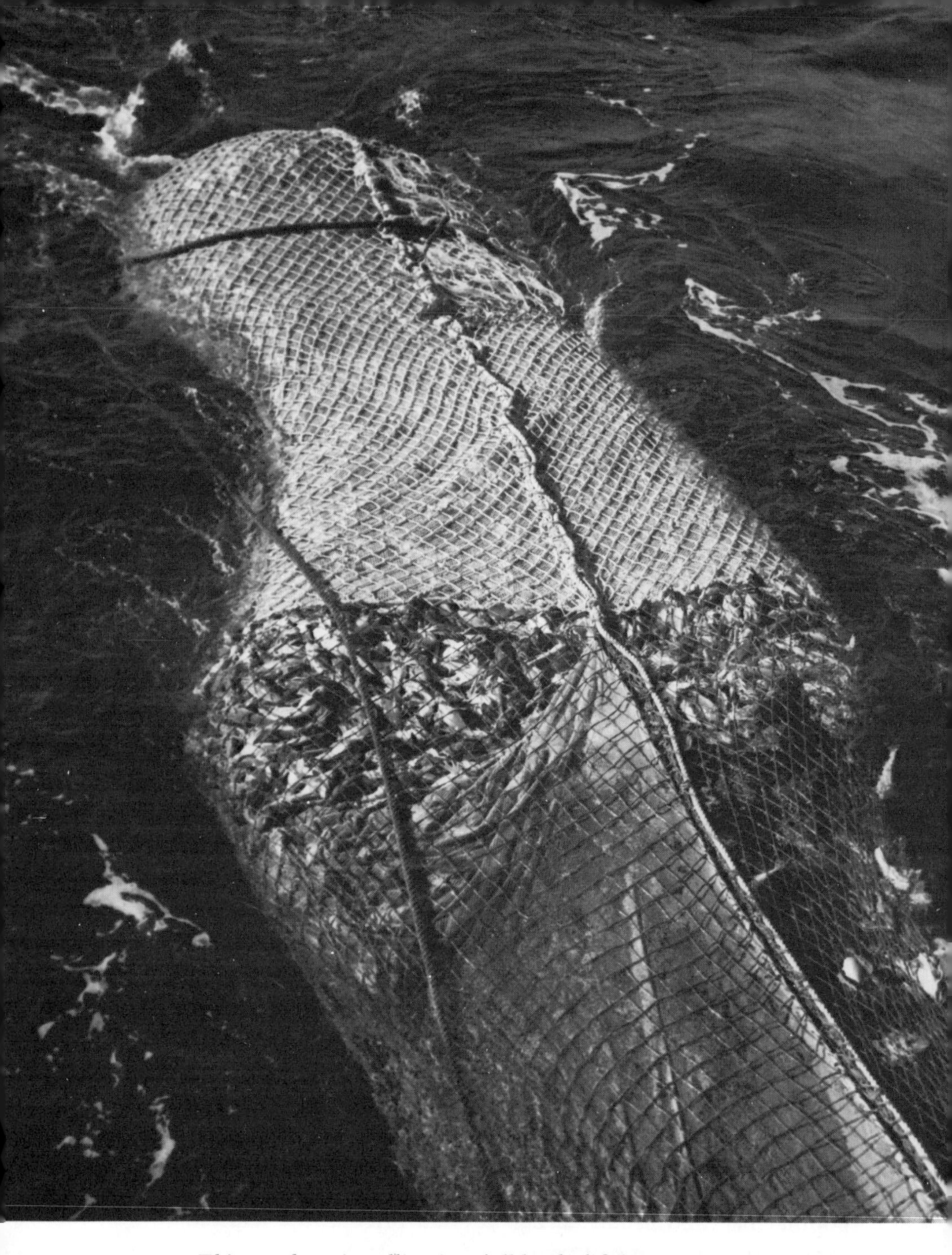

This trawl net is pulling in a full load of fish.
PHOTO: ROBERT K. BRIGHAM, NATIONAL MARINE FISHERIES SERVICE

There are other kinds of fish and marine animals that can be fished and perhaps bring in a profit. There are the so-called trash fish such as dogfish, a small member of the shark family. In the United States this fish is thrown back or ground into fertilizer. In England dogfish is served fried in a fish-and-chips meal and it is possible that dogfish, and other fish that are abundant in the sea, may well be used to take the place of the varieties that are presently being overfished.

There arises still another problem other than overfish and it may be the most serious of them all—pollution. This is the effect of pesticides, foreign substances of any kind and toxic metals that are being washed down freshwater rivers into the estuaries, marshes and coastal waters right onto the shelves. What happens is that some fish are killed outright, some are rendered impotent (producing eggs that are not fertile) and some fish build up a residue of poisons in their flesh, making them unfit for human consumption.

Some great fishing all but disappeared during the sixties. Porgy, fluke, menhaden, spot, weakfish and croaker are some of the varieties of fish that are being seriously affected. It is felt by some scientists that pollution is the major reason for the loss of these fish. The pollutants kill the tiny organisms the fish feed on, as well as killing the young fish and the fish eggs.

Traces of mercury, more than the level recommended by the World Health Organization as a safe level, have been found in mackerel, shad, white perch, striped bass, spotted seatrout, silver perch, false albacore, bluefish, dolphin, whiting, swordfish, tuna and other species. When these fish are caught and eaten the mercury accumulates in the body and, when a certain level is reached, there are problems.

Not all biologists feel that the mercury in ocean fish is a result of pollution. It could be the result of smaller animals picking up tiny amounts in the ocean water and passing it along as they are eaten. Because many fish are at the end of the food chain, they would contain the most mercury. Also it has only been in the past few years that devices were developed for measuring mercury in tissues. The mercury may have been in the tissues for as long as man has been living on earth. Some scientists feel that there is mercury in every living thing.

Residues of the insecticide DDT have also been found in fish that live in the bays, tributaries, estuaries and the waters over the continental shelves. As yet the residues have not reached a dangerous level. But the fact is that all kinds of fish tasted have traces of most of the pesticides that are used or have been used in the past. This cannot be attributed to ordinary amounts usually found in the sea. These chemicals are for the most part man-made.

The old man of the sea has reasons to be sad. The fish are disappearing because of pollutants, overfishing and other man-created upsets of the natural environment. Many people, however, are working to correct these situations so that when the old man of the sea beckons to you to come see something it will be the birth of a billion new fish, not the death of them.

6 Gathering on the Shelves

A land farmer plants his crops, weeds, feeds and waters his plants and, when they are ripe or grown, he harvests what he has sown. That is not the case in the waters over the shelf. There is comparatively little farming of the world's seas, and what there is takes place mostly in saltwater ponds and estuaries. In shelf waters, in the true sense, one gathers instead of harvesting. Most of what man gathers is fish, but there are various kinds of seaweeds gathered as well as many kinds of animals, the most popular being lobsters, clams, crabs, oysters and shrimp.

Gathering seaweed has been going on for about 5,000 years and will probably continue at an ever increasing pace as the years go by. China and Japan, as well as many other countries that have ocean shorelines, have used seaweed as food for centuries. For generations peasants in Brittany, along the coast of France, burned kelp to make iodine and potash. Seaweed also made fine fertilizer, and the Pilgrims discovered that Indians used it for fertilizing their corn.

Later, Robert Koch, a German scientist, used agar, a gel made from seaweed, as a culture medium for the microbes he used in his studies.

Today many varieties of seaweed are gathered. One of the most common is a small variety called Irish moss. It was discovered about six hundred years ago in Ireland and was first gathered on the shores of County Carragheen along the south coast. For many years it was used as a food, fertilizer and medicine. It was boiled with milk, making a setting jelly that was added to various desserts and was the basis for the famous French dessert *blancmange.*

When Scotch and Irish settlers arrived in the United States they brought with them the knowledge of Irish moss. They imported it from home, at a cost of about $2 per pound, but they were more than satisfied to pay the price. But then J. V. C. Smith, who was at one time mayor of Boston, discovered that Irish moss grew in Scituate, Massachusetts, a small community south of Boston. When the citizens of Scituate learned that they could gather seaweed at a profit, they became the first community in the New World to gather the moss.

For years Irish moss was used to heal chapped skin. It became a part of a woman's face cream when it was boiled into a milk gel. Grandmothers and mothers used it to fight the croup and to combat sore throats. Then, in 1888, E. C. Stanford, a British chemist, separated the many parts and named them. The moss extract he named carrageen, because of its ancestry, and it has been called carrageenan (a slight modification) ever since. Later, other areas were uncovered where Irish moss grew in sufficient quantities to be gathered. Rich beds were located along the coast of Maine and in southern Nova Scotia.

The Irish moss grows in the shallow waters of the shelf, on

Raking in Irish moss in the shallow waters of the shelf.
PHOTO: MARINE COLLOIDS, INC.

rock surfaces, held there by holdfasts or stems. The moss is taken by men in dories who use long-handled rakes with long, narrow teeth to rake the moss off the rocks. It is important that the holdfasts be only broken, rather than scraped clean. When the holdfasts are broken, a new head of moss will grow in a few weeks and the same area can be worked again.

The moss is dried and baled, and then shipped to processing plants. One such plant, located in Maine, has machines that extract the two vital products, algin and carrageenan. Both are natural sugars, colorless and tasteless. When added to food, or other products, they act as stabilizers and are often eaten in milk shakes, ice cream, salad dressing, pie fillings, cheese spreads and much more. Carrageenan stops toothpaste from drying out, makes ice cream smoother and keeps salad dressing from separating.

Algin has the remarkable ability to absorb water. Only a tablespoon of algin stirred in a quart of water causes the water to thicken until it becomes like molasses and difficult to pour. Ice cream was the first product to be improved by algin. It gave the ice cream a rich, creamy smoothness.

Algin is used today in other products for the same reason. It can be found in eggnog, salad dressing, cream cheese, canned meat stew, buttered vegetables, chow mein, fruit drinks, fruit syrups, bread batters, chocolate drinks, puddings, pie fillings and chip dips.

Algin also has other properties that offer great value in the field of commerce. It helps to emulsify, stabilize and suspend foods and products, and is also used in making gels and films. It is often mixed in with icing on cake so that, when the cake is packaged for the grocery shelf, the icing does not stick to the box or plastic cover. It is also used in liquid car cleaners, keeping the abrasive suspended throughout the can or bottle. It is added to antibiotics, and to paints to suspend the pigments, allowing the painter to cut down on brush strokes.

It is used in industrial adhesives such as fiberglass mats that are used in storage batteries. Textile manufacturers use algin to keep their dyes from running. The rubber industry uses algin when making synthetic latex. It adds a creaminess to the product. There is even algin in facial beauty masks and hand lotions.

What the moss and other plants furnish is algal hydrocolloids that have great commercial importance. The prefix "hydro" refers to water; the term "colloid" comes from the French language and means "glue-like." Thus, when mixed in water, the hydrocolloid becomes glue-like. There are many varieties of hydrocolloids that come from all kinds of plants on the land and in the sea. There are also some found in animals. Many varieties have various uses according to where they are found, but they do have two things in common—they are versatile and extremely valuable.

Other plants, such as large brown seaweeds, are as important as Irish sea moss. One is kelp, a large seaweed that yields vitamins, potassium and iodine. It is also rich in carbohydrates and algin, and is a source of potash. The seaweed is burned and the potash removed from the ashes. Originally the ashes were called kelp, but later the word "kelp" was applied to the large, brown seaweeds.

Kelp is sometimes referred to as sea otter's cabbage because it comprises one of the main dishes in a sea otter's diet. It is also called bladder kelp, ribbon kelp and bull kelp. In the Pacific it grows along the coasts of the United States and Canada from Los Angeles to the Aleutian Islands. Some kinds grow to be more than 100 feet in length in the Puget Sound, Washington, area. Off the coast of California, in water 25 to 90 feet deep, it has been known to grow to a length of 750 feet. Kelp is attached to the bottom by a holdfast that extends to a whiplike stipe, or stem, which blossoms into brown, wavy blades. It is gathered by using floating barges with a special rig that sheers and gathers kelp 4 feet below the surface. When the conditions are right, the kelp grows 18 inches a day and can be clipped once a month.

The reason kelp is only trimmed and not cut down to the stipe is to preserve the plants. First, this allows the plants to grow better and produce more; and second, and perhaps most important, there are ecological reasons. The many stipes and holdfasts make the surrounding areas a forest beneath the waves on the floor of the shelf. Charles Darwin, in 1834, wrote: "The number of living creatures of all orders whose existence ultimately depends on the kelp is boundless. If any country forest was destroyed, I do not believe nearly so many species of animals would perish as they would from the destruction of kelp."

A holdfast has many different "roots" and holds with it a collection of organisms acting much like a tiny coral reef. Therefore it has been termed necessary by marine scientists that the kelp beds be preserved for the miniature animals that live in the beds and for other animals that feed off the smaller ones.

Similar to Irish moss, kelp is dried before it is sent to market. Once the plant is cut loose, it is sucked up through a pipe to a barge where it is chopped to about the size of a dollar bill and placed to dry in rotating cylinders. Once dried it is bagged and readied for shipment. Five hundred pounds of wet kelp loses enough moisture to weigh only a hundred pounds when dry. If no cutting machines are available, divers work the beds. Some divers can cut about a thousand pounds of this marine plant in an hour. But a diver cannot work an eight-hour day, due to the coldness of the water and fatigue from swimming.

The Norwegian Institute of Seaweed Research has been concerned with seaweed for a different use as well as for the algin content. Seaweed can also be used, in the form of seaweed meal, to supplement the diet of farm animals. It has been determined that the meal can be used to help the coloring of egg yolks and increase the yield of milk in cows. The seaweed meal is made from brown algae in Norway. It is dried in rotary drums and then ground into a fine meal.

Off the southern coast of the United States there is an abundance of turtle grass. This is a flat-leaved plant that has been successfully tested as an additive to animal feeds. It is plentiful. It grows in the Gulf of Mexico, the Caribbean and other tropical waters. In one area off the coast of Florida, an area that covers 4 million square miles, it has been estimated that about 11 million tons of dry leaves could be gathered. But this is only a small area compared to other grass-producing locales in the Caribbean. In tests it was found that supplemental feeding of sheep with turtle grass pellets was beneficial and the experimenters believe that the grass can be used in feeds for other animals as well.

The turtle grass grows so thick in some places that it can be cut and pitchforked into a boat in shallow water. No doubt if the grass proves economical, then cutting machines attached to floating barges could be developed.

These seaweeds do not grow everywhere in the sea. It has been said that the open sea, that part that is not over the continental shelves, is a biological desert. It does not have sufficient nutrients to support the food chain so it produces only a minute amount of the world's catch and has no potential in the future. The shocking thing is that this unproductive area comprises 90 percent of the world's seas.

In contrast, the regions off the west coasts of the major continents and Antarctica produce half of the world's annual sea harvest and comprise only about one-tenth of 1 percent of the ocean surface.

About 60 to 70 million metric tons of fish, animals and plants are taken each year from the sea. Some marine scientists feel that this might be doubled with the most efficient system for taking life from the sea. However, it might be said for man today that his methods of gleaning various sea plants, fish and other marine animals from the ocean is at the same level as Stone Age man in relation to land plant and animal farming. Long ago, man gave up hunting and gathering plants and animals and began to farm them. If he had not turned to farming on land there would not be the vast amount of food available on the earth today. In the sea, man is still a hunter and gatherer for the most part.

Sea farming, mariculture or aquaculture, whatever it may be called, is a practice that has been going on for centuries. But, despite its long existence, it is still a young venture because much of the sea farming has not changed since it was first started so long ago. In the United States, in the early 1970s, there were about 4 million acres of oyster and clam grounds considered to be public or growing naturally. The average yield per year per acre was only about 10 pounds of fresh meat. However, in acres purposely cultivated, about 150,000 of them, the oyster grounds yielded on the average 100 pounds per acre. This was an improvement but compared to France even this was not good.

The French cultivate their oysters using different methods and increase their private shellfish grounds to yield about 1,000 pounds per acre per year. In Australia the method is even more refined and the average per acre is as high as 4,000

These scallop shells are being used to produce oyster seed for commercial use.
PHOTO: MAINE DEPARTMENT OF SEA AND SHORE FISHERIES

pounds each year. The length of maturity is also lowered by a few years. In the United States it takes anywhere from two to five years, while in Australia, the oysters reach marketable size in three years or less. There is hope for more meat per acre in Manila Bay, in the Philippine Islands, where the harvest has been as high as 10,000 pounds per acre a year.

Japan is by far the leader. The Japanese take oyster shells and spread them on the bottom of the bay where oyster spat, the larvae of the oyster, come to rest on the shells. The shells are then strung on wire about a foot apart. The wire is 30 feet long and hung from a bamboo raft that is anchored in deep water. This brings in a yearly yield of 50,000 pounds per acre.

The Spaniards outdo the oyster yields with their mussel culture. Along the Atlantic coast of Spain, in the bays, the Spanish use raft culture to grow mussels. They use ropes that are loosely woven and the mussel larvae are attached to the rope. The rope is suspended from rafts that are placed in areas where there is great tidal change. The mussels grow to marketable size in about twelve to eighteen months. Each raft has about one thousand ropes and can produce 60,000 pounds of mussel meat in a year. With four rafts to an acre it is possible to produce almost a quarter of a million pounds of mussel meat per year on each acre.

The shellfish form of marine sea farming that utilizes rafts has produced the best results so far. Comparing the raft yields to a natural fishery, the best production is off the coast of Peru where the anchovy grounds are estimated to yield about 1,300 pounds per acre per year. In the rich northwest Atlantic fishing grounds, such as Georges Bank, the yearly yield at best is a scant 35 pounds per acre.

With these facts it is tempting for uninformed people to think that by taking so many thousand acres of salt marshes, estuaries or bays that oysters, mussels or any other shellfish

A commercial oyster farm in Oyster Bay, Long Island, New York.
PHOTO: ROBERT K. BRIGHAM, NATIONAL MARINE FISHERIES SERVICE

could be grown by the millions of pounds. But they fail to realize that the rate of growth for these animals is dependent upon how much microscopic food is in the sea water and how fast it gets to the bay, estuary or salt marsh. When there are strong currents the shellfish are usually found on the bottom and they do well. That is where there is the most food. If any form of sea farming is to take place with hopes of a high yield, it will have to be done where there is sufficient food in the water. This is not everywhere and therefore it cannot be assumed that, by merely taking a few thousand acres, sea farmers will automatically produce that much more shellfish.

What the sea farmer must do to help the growth of shellfish is to contribute extensive care. He must lower the numbers of predators in the area and cut down on the other animals that compete for food. This increases the food intake for his particular animal but really does not increase the sea's production. What would increase the production would be to supply the necessary nutrients to increase the phytoplankton. On land this is easily done by fertilizing the soil. In the water there is little, if any, control.

Despite the fact that there is little sea farming going on right now there are many other kinds of animals, other than fish, brought in for food. They come in many sizes, from tiny shrimp to huge whales. Due to their particular availability they are gathered in greater numbers than other sea animals. The catch is measured in tons rather than yield per acre.

The largest worldwide catches of marine animals in the late 1960s and early 1970s included six species of fish and four other kinds of animals. The leader was the Peruvian anchovy, followed by the Atlantic herring, Atlantic cod, mackerel, Alaska walleye pollack and the South African pilchard. The other four animals, all without backbones and referred to as invertebrates, are oysters, squids, shrimps and clams. Of course these catches change from time to time because of overfish and natural causes.

The whale was one of the leaders a few decades ago when many whaling ships set to sea. During this time man sought out and harpooned thousands of whales looking for valuable whale oil. The oil, boiled from the whale's blubber, was needed for lighting lamps. And it brought in a good profit to the ship's owners and captain. The ships sailed from such ports as New Bedford and Nantucket in Massachusetts, and from the coasts of Norway, Scotland and Holland. In the early 1970s only Russia and Japan still hunted whales for food and such by-products as oleomargarine, lipsticks and lubricating oils. Other countries abstained because some whales were disappearing.

During the years when whales were first hunted, it was felt that the whales would last forever. They were so powerful and it took so long to get them, the whales managed to maintain their numbers despite the many years of the kill. But then, as man's tools became more powerful, the plight of the whale increased. Soon larger and faster ships with exploding harpoons mounted in guns in the bow, chased down and slaughtered the huge beasts, sometimes wiping out whole herds, young and old, male and female. As many as 10,000 whales were taken in the Antarctic in 1922 alone. A decade later the take from the same area had reached 40,000 whales in a year. This was the first example of overfish.

There is one animal that is gathered off the shelf floor that is still in fairly good supply. It is the lobster. There are three species of true lobsters including the Norwegian lobster, the common lobster of Europe and the American lobster. There is also the spiny lobster and the crayfish. The American lobster is a very popular item on the menu along the east

Harvesting shrimp from a pond, a form of aquaculture.
PHOTO: LOUISIANA WILDLIFE AND FISHERIES COMMISSION, MARINE LABORATORY

coast of the United States, where the lobster lives. Its range is from the shores of North Carolina to Maine and in Canada as far north as Labrador. It is taken along the shelf in depths of 6 feet to 1,200 feet. There are also some deep-water lobsters.

Three distinct fisheries take place in relation to the American lobster. The first is the inshore fishery where usually one man goes out in his small, thirty-foot lobster boat to haul, empty, bait and reset his string of wooden lobster pots. He usually works in water up to 50 or 100 feet deep and hauls anywhere from 50 to 150 pots a day. He catches small lobsters weighing from about a pound to four pounds in weight. This kind of fishery has existed since the early 1700s and has been profitable, as well as supplied a good-tasting food for the fisherman and his family.

The second fishery is offshore where the water is up to 200 fathoms (1,200 feet). Here large vessels go out from two to ten days to trawl along the bottom with nets for five- to twenty-five-pound lobsters. The fishermen set their nets along the bottom where the lobster lives. If the bottom is smooth, either sand or gravel, then the nets will not be damaged, but the catch may be small. Lobsters are usually found living around rocks, and this becomes a tricky venture for the lobsterman to trawl among the rocks and come out with nets relatively free of tears.

In between the inshore and offshore fisheries is the midshore fishery. Here the bottom is too rocky for trawling with a net, and the depth is too great for the conventional smaller wooden lobster boat. A large metal, plastic-coated trap is used with nylon nets inserted to allow the lobster to crawl in. The lobster boat is more than 100 feet in length, comparable in size to fishing trawlers, and has the latest in navigational equipment. Whereas the inshore lobsterman sets his pots separately, the midshore fisherman strings his traps in a line, 10 to 30 pots tied to the same cable. The cables run anywhere from 100 to 350 fathoms (600 to 2,100 feet) in length and they are fished every other day, although some traps may be pulled every day.

The catch in a midshore lobster pot is much greater than the inshore. The average daily catch for an inshore pot may be from ½ to 1 pound per pot to a high of 2½ pounds. The midshore traps average 5 to 10 pounds per pot per day. The catch is impressive. In a two-year period in the mid '60s, 10 million pounds of lobsters were raised using this method.

Another sea animal collected is abalone. The abalone is a shelled creature that is eaten in soup or as sliced steaks. It is sold both dried and canned. It is Australia's third greatest export from the sea and was worth $4 million a year in the 1970s. At that time the yearly catch was more than 10,000 tons, with four of fifteen kinds bringing in up to twenty-five cents per pound and more.

They are gathered by divers who may go as deep as 75 feet to collect the abalone in nets. A diver dresses in a rubber suit and uses scuba-diving equipment, carrying a steel "ab iron" to pry the animals off the shelf floor. The diver carries a net attached to a line that is connected to the boat on the surface. A net full of abalone weighs about 40 pounds. Some divers earn as much as $400 per day diving along the offshore reefs in deep water.

There are other animals that divers gather off the shelf floor, but here the divers would be happy if they earned as much as $400 in a month. These are the *sphoungarades,* or sponge divers, of Kalymos, a Greek island about five miles wide and ten miles long.

There are three kinds of sponge divers. The *barcas* wear only a bathing suit, going down the same way their ancestors

A large lobster caught in the water over the continental shelf.
PHOTO: ROBERT K. BRIGHAM, NATIONAL MARINE FISHERIES SERVICE

did more than 3,000 years ago. Their exploits were observed by Aristotle (384–322 B.C.) and their feats were recorded in ancient writings.

The *fernezes* wear a steel helmet in what is called hard-hat diving. Air is pumped along a tube that is connected to the hat. The diver breathes in air and is able to stay down longer than the *barcas.* The third kind of diver is the *skafendros.* He wears a helmet plus a full suit to protect him from the cold. Most of the divers are the full-suited kind, and they all live with the same fear, that of a broken air line. If it should break, the diver must use his helmet's safety valve immediately or he will find that the suction created will strip his flesh away.

The other danger is the bends, a form of nitrogen poisoning. Today on the island of Kalymos there are 14,000 residents and 1,500 have been victims of the bends. Some draw a pension from the government, but they must sell fruit, peddle papers or do odd jobs to supplement their income. There is a decompression chamber on the island to aid some victims who are in diving accidents close by.

Once a year the sponge divers leave their small, colorful homes built on hills that overlook the harbor. They head in boats, carrying twelve to twenty men, for such places as Rhodes, Tripoli, Covobuono and Cyrenia in the Mediterranean Sea. They are gone for seven months. They start each day at 4:30 in the morning, with the first diver going over the side shortly after six. As he dives, the boat circles around slowly and the air compressor on board ship toils in a rhythmic thump, slap, thump, slap. While one diver is below, another suits up on deck. Each diver spends about twenty minutes to an hour on the floor of the shelf and makes from three to five dives per day.

When the diver is finished and reaches the ladder of the

A type of glass sponge found along the continental slope.
PHOTO: WOODS HOLE OCEANOGRAPHIC INSTITUTION

boat, his load of sponges is emptied from the net bag onto the deck. The sponges are black and crusted with coral. Crewmen trample them with their bare feet and rinse them with sea water. They are then threaded onto a length of rope and tossed overboard, to be further washed by the sea. Later they are brought on board and scrubbed until clean. Then they are trimmed and put into sacks to be ready for the home port. For all this hard work, the long days, the seven months at sea, the boredom and the exhaustion, the diver can figure to make $1,500. This is less than four days' pay for the Australian abalone diver.

On the Tuamotu Islands in the South Pacific, about 2,000 miles south of Hawaii, there are other divers who go to the shelf floor looking for marine animals. They hope to earn as much as $40 per day. This is not what the abalone diver makes but is certainly more than the sponge diver's income. The divers are after pearl oysters, not for the pearls but for the mother-of-pearl shell. Most divers are naked although some wear a bathing suit. They are also equipped with a pair of tiny goggles and a single glove.

Each small boat has a diver and a helper. The boat is anchored and the diver drops into the water. He hangs onto the outrigger booms and does deep-breathing exercises to build up the oxygen in his lungs and blood. When he feels right, he lets go and sinks rapidly to the bottom by holding a lead weight between his feet. The weight is attached to a line. When the diver reaches the bottom, he drops the weight and looks for shell. Using his gloved hand, he twists the oysters loose from the bottom and puts them into a net that has been lowered by way of another line.

In a short time the diver needs a breath. He grabs the weighted line and starts up hand over hand. His helper on the boat, feeling the diver ascending, pulls on the line to speed up the climb to the surface.

This whole sequence takes about ninety seconds and a diver gathers two or three shells weighing about one kilo (2⅕ pounds) and worth 100 francs (roughly $1.20 to $1.35). The diver rests for five minutes while the helper hauls in the net and weight and adjusts the anchor line so the boat moves to a different location.

By the end of the day a diver will have made between 30 and 40 dives and brought up anywhere from 30 to 80 shells. The chances of a pearl being inside a shell are rare because the divers gather the shell before the oysters reach the age when pearls develop. Despite the lack of pearls the mother-of-pearl shells bring a diver about $2,500 during a three-month season.

The average depth of the dives is 120 feet. It seems like a job for men and it is, but it is not uncommon for two or three women to be seen diving. There are also young boys and many men over sixty. There has been a case of one man diving at the age of seventy-four, and he was apologetic when he admitted he could only go down to 100 feet.

When the divers return to shore, they sell the shell to Chinese pearl merchants. The buyers take the shell to their stores for cleaning and weighing and then payment. The shell is loaded on schooners and sold 300 miles away at a price 50 percent higher than what is paid the divers.

In another part of the world, for 1,500 years, the floors of the shelves off Korea and Japan have been harvested daily by divers, mostly women, called *amas.* Descending to depths of 80 feet by only holding their breaths, they bring up edible seaweeds and shellfish, and sometimes pearl oysters. At one time the main reason these divers went to the floor of the shelf was for pearls, but now their main objective is food.

There are about 30,000 *amas* who work along the seacoasts of Japan and Korea diving up to four hours each day. They dive into the water, take a few breaths of air and swim to the bottom. They can dive repeatedly, spending as much as two minutes beneath the surface before coming up for a rest and more air. Some Koreans dive during the winter but the diving time is cut to a minimum because the water temperature drops to 50°F.

Most of the divers begin when they are eleven or twelve years old and continue to work until they are sixty-five, so it is a full-time, lifelong profession. Women having babies do not miss very much time. They dive right up to the day they deliver and nurse their newborn infants between dives.

Divers work alone or with a helper. Those working alone are called *cachidos.* They have a boat and, after taking several deep breaths, dive to the bottom, usually 15 to 20 feet. Each diver takes up to sixty dives per hour.

The *funada* has a helper. She uses a thirty-pound weight to take her to the bottom and she will be pulled up to the surface by her helper with the rope that is attached to the weight. Using this method the diver dives 60 to 80 feet and is able to spend about thirty seconds on the bottom, twice as long as the *cachido.* Because of the distance traveled, the *funada* makes only thirty dives each hour.

Cuttlefish, squid and octopus are all cephalopods (a class of mollusks) that live on the shelves and are eaten by the Japanese and Koreans as well as by other people of the world. They are not very common in the United States, however. These animals are usually netted. People of south Asia, south Europe and the Pacific islands eat them fried, dried in the sun or boiled. They are sold in the United States in cities where there is a large foreign population such as in San Francisco, New York, Philadelphia and Boston. In Newfoundland and along the coast of northern Europe squids are used for bait in the commercial fishing industry.

Shrimp are also used for bait, but they are better known as

Dredging clams off the shelves.
PHOTO: ROBERT K. BRIGHAM, NATIONAL MARINE FISHERIES SERVICE

a delicious food from the sea. They are netted by fleets of shrimp boats in many areas of the world. Seafood lovers eat them boiled or fried.

Another delicacy from the sea hardly looks edible when first brought into shore. Nothing looks more ugly or more unappetizing than the pincushion of the sea, the sea urchin. But surprisingly enough, these creatures, living on and around rocks found on the floor of the shelf, are gathered by men in boats, brought to shore and sold. The sea urchin is found along the coast of Europe and North America's Pacific and Atlantic coasts. It is an echinoderm and related to starfish, sand dollars, sea cucumbers and others. It eats both animal and plant food, feeding mostly on brown seaweeds. It has many predators including sea otters, sharks, cod, haddock and some sea birds.

The outside spines of the sea urchin are not edible. What is eaten at the table is the reproductive organs, orange and yellow in color. They are eaten raw, usually with a bit of lemon juice. Sometimes the purple body fluid is drunk.

There is a small urchin fishery on the coast of Maine where men in small boats use dip nets with handles of ten feet or more. The sea urchins are scraped from submerged ledges or rocks and, once dislodged, are scooped into the net and brought into the boat. On shore the creatures are packed in bushel baskets and crates for shipment.

Gathering plants and animals from the continental shelves of the world produces a tremendous bounty. It can be a small operation like sponges and sea urchins or something of a much larger magnitude such as lobstering. No matter where on the scale of importance the plant or animal lies, it adds to the production. It is a production that must continue for the sake of the people throughout the world who rely on the shelves for their food and livelihood.

7 Mining of the Shelves

When the old prospector announced one day that there was gold in the hills, he could have also said there was gold in the sea and been absolutely right. There is gold in the sea. In fact there is a fortune of gold, sulphur, silver, copper, salt and much more. The trick is the same today as it was for the old prospector, to get it out.

It has been estimated that a cubic mile of sea water contains 128 million tons of salt, up to 25 tons of gold, up to 45 tons of silver, 7 tons of uranium, over half a million tons of lime (calcium oxide) plus many more minerals in various smaller amounts. But all these minerals exist in such dilute amounts that using today's technology it would be impossible to extract anything except salt economically.

There is, however, some mining taking place on the shelves. In the early 1970s at least 100 underwater mines were active in such countries as Poland, Turkey, Taiwan, Australia, Chile, Canada, Finland, France, Ireland, Greece, Japan, Spain, Britain and the United States. Limestone, coal, iron ore, tin, nickel, salt, copper and gold are the metals brought up.

Coal led the list in value in the '70s with approximately $350 million per year. About $175 million was taken in salt, $150 million in sand and gravel and $75 million in magnesium metal. Tin and aluminum are found in scattered areas with traces of gemstones in Cambodia, South Vietnam and Thailand, and bits of gold in the Gulf of Thailand.

Metals with such strange sounding names as zirconium, thorium and titanium are brought in from the beach and sand dunes in parts of the United States and in Africa, India, Ceylon, Australia and New Zealand. These metals are used as foundry sands and in the aerospace industry. There is a big demand for them, and their exploration and production is expected to increase from high-grade shallow-water deposits.

The idea of mining the sea and the shelf floor has some good points compared to land mining. When minerals are mined on land they are not replaced. The ore is used up and then the search for more ore continues, making the mining more expensive and more difficult. Because the sea keeps receiving the deposits as a result of runoff, erosion and decay, the sea is a reservoir and thus is able to replenish its supply. It takes thousands of years to build up a thin layer of manganese, for example, but if and when man does begin to mine the sea of minerals in any great quantity it is believed the supply will not easily be exhausted.

Today there is a mining venture taking place right on the continental shelf that all of us can recognize. It is the sand and gravel industry. This material is under increasing demand in the construction trade as fill, and for concrete and road beds. It is also used to replenish coastal beaches that have been washed away by waves.

Samples taken from the outer continental shelf show that sand and gravel far offshore are coarser than the sands found near shore. The sand is usually brown and stained with iron, giving scientists the impression that it has been exposed on the bottom for a very long time.

A study by the Geological Survey of the United States Government, made in the late 1960s, indicated that most of the shelf off the coast of the northeastern United States is covered with very fine-grained to coarse-grained sand. The sand is many feet deep, the exact thickness not determined. It is thought that the sand could be profitably mined since it is close to metropolitan areas such as Boston, Providence and New York.

It is also believed that mining sand from the shelf must be done because land deposits of sand and gravel are being used up at a rate of 6 percent per year. Also the high cost of handling and transport makes it impracticable to ship sand and gravel over long distances. Zoning laws make it impossible for sand and gravel pits to operate in some communities and others have been closed down. Many gravel pits have become more valuable as housing sites so owners have no choice but to take the money when offered. These losses of sand and gravel sites on land suddenly make the shelf deposits more economical.

Sand is cheap in the sea and also easy to mine. The sand is pumped up from the shelf and piped aboard barges. If the need arises to store the sand, it can be dumped in shallow water and stored there until needed. There are also dredges. One is the self-propelled hopper type that is used by the U.S. Army Corps of Engineers to rebuild eroded beaches. This type of dredge can also be used for mining sand and gravel.

The underwater sand and gravel seems to have some

advantages over the sand and gravel mined on land. When it is mixed with asphalt and similar materials the salt crystals that are attached to the particles of sand help to prevent ice heaves in winter and absorb expansion caused by heat in summer. Of course the salt does not remain in the sand for the lifetime of the road but it is there until it seeps out.

A similar substance may be taken from the shelf floor in the future. It is mud. It is thick, black and has a high silt clay content, and in tests it was found to be ideal as economical building blocks. The mud is found in areas off the northeast and Pacific northwest coasts of the United States, off the coasts of England and also in the North Sea. The mud is a direct result of the ice age glaciers that left the mud on the shorelines.

The mud is in depths of 40 to 70 feet, several miles off shore, making it a good distance from shore beaches and wetlands. On the shelf floor it is sticky and black but, when dried, it is gray and brittle. When fired (hardened by heat) the brick turns red and is stronger than many bulk building materials in use today.

If mined, the mud would be taken at least a foot below the shelf floor, leaving the top twelve inches. The pure mud contains about 30 percent water, 10 to 15 percent feldspar, 10 to 30 percent quartz and 20 to 60 percent various kinds of clay. Some of the mud has up to 10 percent calcium carbonate derived from shell. Since there is a low percentage of organic matter, it is especially suited as a building material.

Scientists involved in studies of the mud feel it would be sound, ecologically, to mine the mud because what few mud-dwelling animals and plants found there would be able to reestablish themselves. What would have to occur is

This cliff, showing layers of clay and sand, was underwater 20 million years ago.
PHOTO: U.S. DEPARTMENT OF THE INTERIOR, GEOLOGICAL SURVEY

mining confined to small areas and periodic changes in mining sites to allow the dredged areas to repopulate.

Among the difficulties that occur when mining mud, sand or gravel from the shelf are the problems of who owns what. Some countries have no such problems, as they are isolated and no one shares their borders or shelves. Other countries are crowded together and often share the waters above the shelves. It will have to be resolved who has the rights to mine what areas of the shelf and for how long. These international problems are worked on continuously, and most likely it will take years for the task to be fully accomplished.

The shelf area off the United States coast is owned either by the states or by the federal government. From shore to three miles out, the states have control. From three miles to the edge of the continental slope, the federal government has control. In order to mine on the shelf, mining permits must be obtained from the state or federal government. They offer leases.

Conservationists and some marine biologists do not like the idea of disturbing the ocean bottom, as they are not sure what might happen to animal and plant life. They believe that pumping sand from the shelf floor might cause a change in the amount of free-floating sediment, which might alter life patterns of migratory fish. Storing sand in shallow water would bury whatever life was there. The creatures that get uprooted in the dredged areas might cause a break in the food chain. Shellfish grounds might be destroyed and the feeding grounds of sportfish and food fish might be ruined, causing the fish to seek food elsewhere. Mining might kill animals in numbers that can never be counted. Everyone does agree, however, that adequate research must be done before anyone disturbs the shelf floor on a large scale.

Still tests continue. Prospectors today are looking for

gold, and other valuable minerals, in the Alaskan seas, in the sediment located on the continental shelf and on the bottom of bays and fjords. Other minerals the prospectors hope to find are rutile (the ore of titanium), platinum and silver.

There is a reason for looking for gold in these particular areas. All during the ice ages, and up to 200 years ago, thick ice stretched from the sea all the way back to the Alaskan mountains. Slowly these glaciers edged down the mountain slopes, tearing away the terrain as they moved. The rocks were ground into sand and minerals went, too. As the glaciers melted, the water ran toward the sea taking along the sand and minerals. If and when an economical means is developed to bring up the sand and sift out the minerals, such as gold, then this area may prove to be valuable.

On the other side of the North American continent, in the Atlantic Ocean, a venture took place off the coast of Nova Scotia. It is interesting to follow one method of exploration to see how it is done. A research expedition during the 1960s looked for gold on the shelves because it was decided that, if there were onshore gold deposits nearby, there should also be offshore areas with more gold. There were many geological similarities between the sea and land deposits to make the scientists think an experiment was feasible.

Sites were selected for study on the basis of whether the geological factors indicated a good chance of finding a gold lode deposit. When a rich-looking area was found, time was spent outlining the past and present features of the shelf bottom to locate the best drill targets. What the scientists looked for were buried stream channels, ancient beach lines or scour basins, any place that indicated there was water action that would separate and concentrate the heavy minerals. Side surveys also included looking for products such as sand and gravel, phosphate for fertilizers and even clay for ceramics.

Almost 200 drills were made in the prime areas. The drill used was a combination of a vacuuming pipe for lifting and a jet-stream pipe for cutting into the sediment. Hoses were attached to the pipes to convey the sediment to the ships above. The material that was brought to the ship was broken up and mixed with water as it passed over a sluice box. This extracted the gold and other heavy minerals. Targets that were drilled first brought up only traces of gold. More drills were made. As the prospectors zeroed in on the stream bed or channel, they came up with deposits that yielded greater amounts of gold. Values in the first drill area ranged from a few cents of gold per cubic yard to over four dollars per cubic yard, based on the price of gold per ounce at that time. The gold deposit was estimated to have 6 million cubic yards of gold-bearing sediment.

Another deposit was drilled three miles from shore. A remnant of a hill or small mountain that had its top eroded away was discovered. The mountain was found by seismic tracings of the sea floor, using the method of sending pinging sounds from a ship to the sea floor, and then reading the echoes. It was suggested that, as the top of the mountain had been worn away by weather and glaciers, the heavy metals, which were difficult to wash away, remained behind. Many holes were drilled and in each a sample of coarse gold was found. The worth was estimated to be about 30 cents per cubic yard, located in water from 85 to 120 feet deep. The mountain contained an estimated 36 million cubic yards of gold-bearing sediment. Despite the low value of the find, the surveyors felt that because so many coarse gold particles were brought up, a high-yield source was on the bottom somewhere nearby.

When prospecting for gold at sea, it is not only important to learn the layout of the continental shelf but it is equally important to read correctly the samples that are brought up. Tiny particles of gold called "colors" can be found long distances from the main source either on land or in the sea. These tiny specks or fine-grained pieces are very thin and flat, and can be carried by the water action much more easily than heavy particles. Gold particles on land are freed from their rocks by chemical weathering and the outside forces of wind, rain and ground movement. Heavier gold particles lag behind the rock particles, staying with the coarse-grained sand and gravel. When a prospector spies the coarse sand and gravel, he or she knows that gold must be in there, too. What the oceanographic prospectors must do is to find out from what direction the colors came so that they can trace back and find the coarse particles of gold and perhaps the main source.

During the drilling experiments off the coast of Nova Scotia there was often poor weather or high seas to interfere with the operation, pointing up some of the many hazards of drilling at sea. But there is gold in the shelf bottom. A way must be found to get it up that is practical and economical. Also there must be a determination, as there must be with the sand and gravel mining, as to the possible extent of damage to the ocean environment. When those two factors have been resolved, there may be a day when some gold will be mined from shelves around the world.

A different kind of mining is taking place off the coast of Florida near the edge of the Great Bahamas Bank. A man-made island has been created south of Bimini, one of the Bahama Islands. The reason for the activity is simple. Unlike most ocean mining, there is money to be made because of the creation of a unique deposit.

For the past 5,000 years, water from the deeper bottom has flowed across a spot called the Bahamian Shallows. As the water passed over, the sun warmed it, causing calcium carbonate to precipitate out of the water. As a result there are long bars and broad drifts of pure argonite, an ingredient that is used as a soil neutralizer and in the making of cement.

On land, limestone, which is the major land source of calcium carbonate, must first be dug from a quarry, crushed and crushed again and perhaps refined before it can be used. The argonite in the Bahamian Shallows is already a very fine grain and quite pure. Therefore it can be used in agriculture or industry with little preparation. The total drilling area is more than 8,000 square miles and, figuring rock-bottom prices, the whole deposit is worth about $15 billion.

However, the problem is similar to that of the sand and gravel operation. The argonite can be sucked up easily enough and piled on the man-made island, but can it be shipped to market cheaply enough to compete with land deposits? Some people think it can.

Again conservationists fear the worst for the dredging operation. A huge hydraulic dredge can pick up 10,000 cubic yards of argonite each day. As it does, it also sucks in about 10 million gallons of water. The result is slurry, the suspension of fine particles of argonite in water. What forms is a white cloud in the water that could drift for miles. Since dredging usually takes place every day, twenty-four hours a day, this white cloud could indeed grow larger and larger until it covered many square miles of shelf water. Conservationists want to know what happens to the environment when such a large cloud is created.

It is realized, however, that clouds have been forming in the area through natural causes such as hurricanes. These clouds have lasted as long as a week and the life in the shallow water has been able to survive. Nevertheless, the man-made cloud may be the final touch that tips the ecological scale and causes widespread havoc in marine life. There is much turtle grass in the area and many coral reefs. The turtle grass is loaded with life, tiny organisms that feed on smaller organisms and in turn are eaten by larger animals. In a study of turtle grass, it was learned that a little more than a square yard of turtle grass contained more than 28,000 mollusks such as clams. In another study, involving the kinds of fish that live above the turtle grass, more than 100 species of fish were found.

In the clear water around the Bahamas the turtle grass grows to depths of up to 40 feet. The clear water allows the sunlight to reach that depth. In waters around Biscayne Bay, near Miami, Florida, turtle grass does not grow much deeper than 10 feet. The reason is the lack of light in the turbid waters. Because of this conservationists predict that the white cloud, if allowed to form, would seriously damage the turtle grass and, in so doing, wipe out a good share of the life that lives among the grasses.

The coral reef, at best a very delicately balanced community, could be in trouble because of drifting silt. When silt particles drift down and settle on the coral, tiny coral polyps attempt to rid the reef of the silt. If the job is too much, the polyps retreat and wait. If the silt is too thick they wait so long that they die. When the polyps die the reef begins to break up. If there is no silt, but just cloudy water, the lack of light also affects the reef. The coral attempts to build at a deeper level and often does not survive at such depths.

Similar situations occur throughout the world wherever mining on the shelf is underway in a large operation. No matter what takes place, dredging, drilling or explosive surveys, there is damage to the environment. In many cases

Branch coral growing at a depth of 900 feet.
PHOTO: WOODS HOLE OCEANOGRAPHIC INSTITUTION

the damage cannot be measured and, in some places far away from population, the dredging takes place with little, if any, interest from anyone except those who try to make money from the sea and shelves.

Among the most often mentioned minerals to be taken off the shelf as a money-maker are manganese nodules. They are found in many areas throughout the sea and vary from light concentrations to very heavy concentrations. Investigators feel that the nodules are not worth mining, however, for the manganese alone. They may be valuable for the nickel, copper and cobalt that are also in the nodules. Copper has been considered to be the most profitable. As yet, it has not been feasible to mine these nodules in great quantities because of the depths, 1,000 to 3,000 feet and more. Again the economics of the mining is the major drawback.

Other minerals that are often mentioned as mining possibilities are phosphorite, tin, iron and diamonds. Diamonds are being mined off the coast of South Africa, but the cost of mining them often exceeds the worth of the diamonds. Tin is mined in the shallow waters off the coasts of Indonesia, Thailand and Malaysia and a little bit of iron is mined off the coast of Japan. But the Japanese realized it was more economical to bring iron in by ship from Australia than to mine it right off its doorstep in the sea. Phosphorite is also found in many areas along the continental shelf and slope. Because the value of phosphorite is less than $20 per ton, it does not look profitable. And no one is going to mine any of the products from the shelf and lose money doing so. It is hoped that many of the minerals may be worth mining at a later date—if and when technology is more advanced and the prices rise because of scarcity of resources on land.

Sulphur mines do work at a profit and operate off the coast of Louisiana over several hundred acres. This is the third largest sulphur deposit ever found in the United States. The sulphur is mined the same way as on land. Hot water is pumped into the mine to melt the sulphur. The melted sulphur is then aerated with compressed air and the mixture is forced to the surface. About four-fifths of the sulphur mined in the world today is used in sulphuric acid. The acid is used in the manufacture and production of detergents, paper, drugs, petroleum and rubber.

With all that man has taken from the sea and all that he plans to take in the future, the most valuable item, the one thing man depends on most for life itself, is fresh water. Fresh water is taken from salt water over the shelves. It is well known what the importance of fresh water is to the survival of life on earth. And the oceans have an inexhaustible supply. The supply on land is not adequate. Each summer it is commonplace to read that somewhere in the world a drought has caused cities and towns to ration fresh water. The end of droughts may not occur but the shortages can be helped with the building of desalination plants.

In the 1960s many plants were erected when a process was finally introduced that cut the cost of separating fresh water from sea water, thereby making it an economic possibility. Before this, for example, fresh water was piped from the mainland to Key West, off the coast of Florida. Today Key West gets its fresh water from the shelf water that surrounds it.

One of the most ideal uses of the shelf is to use tidal energy for power. For years men have watched the tremendous power of the sea occurring above the shelves, in the rise and fall of the tides. If this power could only be harnessed, then man would have all the energy he would ever need. Imagine building a plant right on the continental shelf and having ocean water run through giant turbines, generating electric-

ity. Because of increasing demands for new energy this is not a dream anymore, it is a reality.

Years ago water mills were built on streams and rivers to harness the swift currents and turn them into energy to run machines. Two and one-half centuries ago Bernard Forest de Belidor expressed in a treatise the idea of putting the tides to work on the tidal mills located in Dunkirk, on the coast of France. These thoughts interested inventors but nothing came of them and they were lost. But, in 1959, a project was designed to harness the tides. This plant, located on the Rance River, in France, along the English Channel, was begun in 1962 and finished in 1966. It is a complete power plant utilizing the energy from the tides.

The French project is essentially a dam that completely cuts across an estuary. It cost more than $100 million to build, and two million cubic yards of water had to be removed after a series of cofferdams were built, drying out a total of 190 acres of the estuary. Today, when water passes the turbines, electrical energy is created. At first it was estimated that the project was too expensive and would take seventy-five years to pay for the investment; however, with prices of oil and other energy rising, tidal power plants look better and better.

In order to build other plants, a substantial tide drop is needed. The greater the tide drop, the greater the energy potential. Turbines are able to turn when the basin behind the dam is filling up on an incoming tide and when the basin empties on an outgoing tide.

There are locations similar to the Rance River, throughout the world, that are well suited for harnessing tidal energy. Across the Channel in Wales, Scotland and England, thirty-two locations have been deemed available for the establishment of tidal power plants. Twelve of the locations have a tide drop of more than 33 feet. France has thirty-nine locations and the Seoul River, in Korea, has differences in tides of 43 feet as does one in India, in the Gulf of Cambay. Australia has two excellent locations with tide drops of over 33 feet.

Five locations have been found in Canada, one in Anchorage, Alaska, one in the mainland United States, one in Mexico, five in Argentina and one in Brazil. All told there are close to one hundred sites in the world that are suitable for tidal plants, with some of the sites close enough to be coupled together into one plant for greater efficiency.

Some of these sites are under construction but no one is rushing into the venture yet. There is the possibility that nuclear power will be cheaper than harnessing the tides. At any rate, tidal energy is just one more in a long list of uses and products that come from the sea over the continental shelves.

8 Oil and Gas from the Shelves

In some countries of the world a boat skipper may cast his lines from the dock, take a deep breath of sweet salt air and then head out to sea. On the way he may have to dodge a multitude of three-legged platforms that stand in the water like giant prehistoric birds, looking as though they are waiting for a huge fish to come by so they can gobble it up. These are not birds but places where men live and work to drill oil and natural gas from the floors of the continental shelves.

The continental shelves are practically oozing in potential oil, according to oceanographers. Off the coasts of South Vietnam and Cambodia are 400,000 cubic miles of oil-bearing sediments. There are also suspected deposits in Southeast Asia, in the South and East China seas, the Yellow Sea, the Gulf of Siam, and in the Andaman and Java seas. Between Burma and Korea are potential oil deposits in an area comprising more than a million cubic miles, making this one of the largest untapped petroleum sites in the world.

Along the east coast of the United States is Georges Bank, considered to be a major deposit area, and further south there is the Baltimore Canyon Basin, off Delaware and New Jersey, and the Blake Plateau, off northern Florida. A secondary field is just to the north of the Blake Plateau and it is called the Georgia Embayment, extending from South Carolina to mid-Florida.

The southern coast of the United States, in the Gulf of Mexico, is dotted with thousands of wells and more submerged land is being leased to oil companies so they can do more in the way of exploration. On the west coast of the United States, oil platforms are freckles on the waters of the continental shelves off southern California.

Britain and Europe are tapping the North Sea for oil, hoping to brighten their future with light, heat and wealth. In the 1960s there were no wells there but, by the early to mid-'70s, nine oil fields and two natural gas fields, both major, had been located near Norway. Three fields were found off Denmark and seven gas and one oil field were located in Dutch waters.

During this time period only 50,000 barrels of oil per day were produced from the North Sea fields—hardly more than a drop in the proverbial bucket. But production is expected to increase rapidly. By 1980 3 to 5 million barrels of oil per day will be brought up from the North Sea, including oil from the Argyll Field located 200 miles southeast of Aberdeen, Scotland. This is Great Britain's oil field of the future, according to some experts, because if this figure is true then Great Britain will become one of the world's great oil producers.

Major oil companies working the North Sea fields are experiencing difficulties as they search for the expected total of 40 billion barrels, more than the entire expected total reserves of the United States. Winter winds reach 100 miles per hour and mammoth waves, some 80 feet high, have crushed semisubmersible drilling rigs and buried them in the sea. But the hunt continues and millions of barrels of oil will be taken from this shelf someday.

Venezuela has one oil base, Lake Maracaibo, that is sprinkled with oil derricks, looking like a forest with a watery floor. Oil has made this Latin American country one of the richest, if not the richest, Latin nation.

Productive shelf waters can make nations rich. But with the riches comes trouble. These same shelf waters also receive the greatest influx of oil, usually in the form of spills, and this creates one of the major problems of this era. Oil as an energy form is needed, desperately needed, but what is not wanted is the soilage that unfortunately seems to go hand in hand with the oil.

Oilmen, lawmakers and conservationists are constantly at odds with each other over the newly created practice of taking oil and gas from the shelves. Until a new energy source is found, one that can meet great demands and be competitive in price, the once untouched shelf floors are going to be drilled to extract the precious oil and gas. And conservationists are going to say that the oilmen are killing the wildlife and destroying the beaches. Oilmen will counter that they are not destroying but are, in fact, doing much more good than harm. Like anyone else the lawmaker, caught between this squeeze, finds it difficult making the correct decision. He or she has to determine what is fact and what is fiction, and that is where the difficulty begins.

To see what problems the lawmakers face, it is best to detail some background information on oil and its uses. There is mention in historical records that the Arabs and Persians, around 3000 B.C., used oil for medicinal purposes, for war and for light. Alexander the Great had flame throwers that roared because of oil. In fact, around the year 1200 A.D., the Chinese had drilled 3,000 feet into the earth in search of oil. About the same time Marco Polo noted that there was a working oil business in Persia. Oil today is needed to wage war and to help keep the peace. Thousands of items are made of oil or are directly related to oil and its by-products.

The tremendous growth of the world population in the last few decades has resulted in a staggering rise in the use of natural resources. In one glaring example, the United States has consumed more natural mineral resources since the turn of the last century than the entire human race used in the previous several thousand years. There is a great concern as to how and where all this demand for energy will be met. Someday it may all come from nuclear power, the sun or other sources but, until that time, it is felt that the energy needed to perform all the requirements of modern civilization will, in increasing portions, come from the continental shelves.

It has just been during the past quarter of a century or so that man has taken oil and gas from the sea floor. Estimates have been made that the potential oil fields located on the continental shelves of the world may be in the trillions of barrels, which is equal to the estimated total for land reserves of liquid petroleum.

To drill an oil well from a platform suspended over the water costs more than it does to drill a single well on land. But oil companies go ahead because they have found that drilling on the shelf produces a better ratio of oil than for wells drilled on land. They have more hits on the shelves, which means a higher batting average. Therefore oil companies have spent billions of dollars in royalties, lease costs and drilling to develop an industry that is suspended above the waves and often many miles from dry land.

In the early 1970s approximately 15 to 20 percent of the world's production of oil came from the offshore drilling sites in the waters of more than thirty nations. More than fifty nations were engaged in surveys. The major areas were located off the shores of Alaska, Canada, California, Central and South America, the Middle East, the Gulf of Mexico and the North Sea. There were many other spots under exploration and many will be developed as soon as the drilling rigs are towed into position. This illustrates that shelf drilling is growing rapidly.

It grows because the need for oil increases. It was estimated in the late 1960s that a billion barrels a year would be needed above and beyond what was already being used. The result was that the oil industry tried to find areas that met this billion-barrel demand every year just to stay even. What this actually suggested was that the oil industry must produce as much oil and natural gas in the next fifteen years or so, as it did during the first century since the first oil well was drilled in the year 1859.

The investment in offshore drilling and exploration is staggering. It is $20 billion to $30 billion and going up. Offshore drilling has increased by about 20 to 25 percent a year and the future promises that it will continue to increase, with no immediate end in sight. Oilmen realize that 90 percent of the good dry-land sites have been drilled. Many are still pumping and bringing up oil, but the wells will go dry someday. That will leave only shale and the sea as sources of oil. Shale has not been proven economically and environmentally sound. On the shelves only 10 percent of the potential shelf sites have been explored. This leaves a vast area still untouched and this causes oilmen to pace the floor in anxious moments when they think of the oil that awaits discovery.

Drilling offshore has its problems. It is expensive, about three times as much as drilling on land. Drilling costs overall have also gone up in recent years and there is every reason to assume that, like anything else, the costs will not stop rising. Oil prices also fluctuate, leaving the oilmen with a question. At what depth does it become impractical to drill? This would be the depth where it would cost more money to bring up the oil than it would bring in when it is sold. And no one wants to lose money. The question remains unanswered. As the price of oil rises, the depth oilmen can drill increases, provided drilling costs do not wipe out the price increase.

The great quantities and prices of oil mean that the oil and gas industry is responsible for more than 90 percent of the value of all minerals taken from the oceans. There is also the greatest potential in this industry. It has been offered that, by 1980, about a third of the oil produced in the world will come from the shelves. That is four times the 6.5 million barrels a day that was produced in 1970. About 10 to 12 percent of the gas production was predicted to come from the shelves by 1980, compared to the 6 percent in 1970.

Since 1946 there have been more than 10,000 wells drilled off the coasts of the United States. Up to 1970 more than six million acres of the outer continental shelf off the United States had been leased for oil exploration and survey. The lease income to the federal government was almost $3.5 billion.

The leasing continued. In the 1970s the United States Department of the Interior, for example, leased an additional 600,000 acres, representing 127 tracts, off the coast of Louisiana. The leases were sold to the highest bidders. The reason was that the oil and gas that might be found there

A drilling platform at sea.
PHOTO: EXXON CORPORATION

would be needed to avert a national energy crisis or at least slow one down. In turn the leases provided more income for the federal government. In 1973, 147 tracts were leased off the shores of Florida, Mississippi and Alabama.

How does the oil get off the shelves and into cars, homes and industry? In 1969 it was estimated that at least 175 different kinds of oil rigs, with platforms of football-field size, were located on the shelves. Of all these kinds of platforms there are five main types called drilling rigs or units. The first is the submersible that supports a drilling platform on columns that are attached to a hull. The whole unit is made on shore and towed to location by tugboat. The hull is filled with water and sunk, while the platform remains above water. When the well is completed the hull is pumped out and refloated, and the whole rig is towed to a new location. The submersible platform is considered to be very stable, but is only used in depths of 100 feet or less.

The jackup rig, a second type, is used in depths of up to 120 feet. It is similar to the submersible since it is built on shore and towed to the drilling site. In this case, however, the platform is floating with legs sticking up like an upside-down table. Once at the site, the legs are lowered to the shelf bottom and the hull is jacked up to the desired level above water. After the well is drilled, the legs are raised, the platform is lowered and the rig is towed to another location. The major disadvantage of the jackup rig is that it can only be towed at a rate of two knots (2.3 miles) per hour, making it a long, tedious haul between drilling sites.

A third, the floating-type rig, is self-propelled. It is in the conventional shape of a ship, with bow and stern, and has the platform located on top of the hull. Outboard anchors keep the rig in position over the drilling site. It is not as stable as the other two types because it is not sitting directly on the shelf. But it is more mobile and can be used in depths of up to 900 feet.

The semisubmersible, the fourth type, is operational while it is resting on the bottom, floating or partially submerged. It is moored in position with anchors. It can be used to depths of 900 feet but must be towed between drilling operations; because it is a cumbersome rig, the towing is a slow process.

The fifth and newest type of major drilling rigs is the anchorless-position-keeping unit. What is new is its use of the concept of dynamic positioning to keep its position. The drilling rig is built onto the ship's hull and is able to move with some speed to the drilling site. By using computers the ship is kept in position, controlled by automation, similar to an automatic pilot. The computer, allowing for currents and waves, keeps the drill in alignment. This rig is capable of drilling into the shelf or sea bottom to any depth but, when used for oil wells, the limit is 1,200 feet.

A typical platform may have forty-two men who work twelve-hour shifts each day. It is about the height of a twenty-story apartment building and could be located 100 miles offshore and be drawing oil from 7 to 13 big oil reservoirs that are from 5,000 to 15,000 feet below the ooze of the shelf floor. Each day 28,000 barrels of crude oil can be pumped to the mainland, seven days a week, through a seven-inch pipeline.

On the platform oil and gas go through tubes into gas and oil separators and then out a pipeline that connects the platform with the mainland. The wells branch out in sharp angles similar to the legs of a giant daddy longlegs. Some travel down and out, tapping oil a mile from the legs of the platform.

During drilling each bit rotates at 140 revolutions per minute and the rig can push up to 250,000 pounds of pipe

The oil-drilling ship Glomar Grand Isle.
PHOTO: EXXON CORPORATION

with a thrust of more than two tons. The rig beats and throbs every hour of the day, shaking the concrete bunkers that act as home for the men during their stay on the platform. Some of the men work shifts of fourteen days on and seven days off. It is a very tiring job but, as some of the men have little education and come from poverty areas, they realize that the back-breaking work pays much better than what they might earn at another job.

The men who steer the bit during the drilling work an equivalent of six months of the year and average more than $25,000 for their efforts. The less skilled laborers and roughnecks, working the same six months, would earn half that amount. During the drilling operation the men haul up 10,000 to 14,000 feet of pipe every sixteen hours to replace the bit. The hauling takes about seven hours but it takes only seven minutes to change the $300 bit.

In 1970 there were more than 6,500 platforms in the Gulf of Mexico. New platforms were being added at the rate of 400 per year and the total was climbing. With all those platforms and the miles and miles of pipe, the danger of an oil leak or a blowout increased. Oil leaks have happened, causing concern among a great many people. During the late 1960s more than 3 million gallons of oil leaked from a well, off the California coast, in the Santa Barbara Channel. The oil drifted to shore and caused much damage but was also instrumental in establishing many of the safety measures on the drilling and piping operations off the coasts today.

The shelf waters, the most productive areas of the sea, receive the majority of oil spills and destruction. The first major oil spill to receive worldwide attention occurred on March 18, 1967, when the tanker *Torrey Canyon*, carrying 117,000 tons of oil, was wrecked in the English Channel. It was estimated the spill was about 30 million gallons. Up to that time mankind had never experienced an oil spill of that magnitude. Consequently, those that were charged to clean up the spill made errors.

Detergents were used on many of the rocky Cornish beaches and coastal areas to get rid of the oil as it came ashore. But the detergents did more damage to the marine life than the oil had. The detergents killed off all the animal life and most of the plant life and sometimes formed patches that lasted up to twenty-four hours. When the patches were pushed out to sea, they were blown ashore in other areas and did more killing. The French used less detergent and also used chalk to sink the oil. This method seemed to work better. There was less loss of life, but it was never determined what damage might have been done to the ocean bottom where the oil accumulated.

The beaches were left sticky, black and smelly. It was a tremendous effort to clean up the shorelines and it was many months before the coast seemed to even hint of returning to any kind of normality.

At Santa Barbara the beaches were also ruined and boats in marinas became covered with black, sticky ooze. There was also a strong odor and the people who lived along the coast were aroused indeed.

Sea birds were covered with oil and died. Seals and other larger animals could not find food and died in quantities that broke the hearts of conservationists.

For years, scientists had been studying the effect of oil spills on wildlife and the environment in general. The major oil spills caused other scientists to take a more direct look. They realized that oil pollution was inevitable when the whole world used oil as a basic source of energy. Therefore the use of oil without accident or spillage is just about impossible. There are losses during production, in transporting the oil, in refining it and then in its use. And the same is true for many of its by-products.

In order to determine the approximate amount of oil in the ocean attributable to spills, figures were obtained from the British oil port of Milford Haven. During one year 2,900 tons of oil were spilled or lost. This represents 0.1 percent of the total of 30 million tons that went through the port during the same year. The losses resulted from breakage, faults in design, mechanical breakdowns, poor human judgment, and there were losses during transfer. There were also losses such as accidents in shipping and flushing of bilges, where oil barges were often cleaned of sludge at sea. After determining the amount of worldwide oil production, the amount of oil spilled could be assumed to be about 0.1 percent of that total—which is about a million metric tons. The actual amount of oil spread across the seas would be higher since the figures do not include spills from oil rigs on the shelf or the return of oil products from runoff of streams, rivers and disposal wastes. There is also residue from marine fuels that do not burn completely. With all this oil, added to the million metric tons spilled, the total could be at least 10 million tons and probably much, much, more in the sea today.

The kind of oil spilled creates different problems. Crude oil has many parts and these different parts have various boiling points. When crude oil is heated to higher and higher temperatures, the various parts boil off. These parts are such well known items as gasoline, kerosene and lubricating oils such as those used in automobile engines. Many of these parts contain types of poisons that kill when they come in contact with, or are ingested by, animal and plant life.

Just what the long-term damage is to the environment from oil spills is not clearly understood. Preliminary studies revealed that when an oil spill occurred particles of oil eventually broke up into tiny droplets. Then these droplets were absorbed by many small animals and plants that live in the sea. Like pesticides, the oil remains in the systems of the small plants and animals. It is then passed along the food chain, from tiny animal or plant to its predator on up the chain and, eventually, to the animals that are eaten by man, such as shellfish, lobsters, fish and even whales. What happens is that the food may taste bad or, even far worse, there could be an accumulation of poisons or even cancer-causing compounds.

At first it was also believed that if some animals did not die when subjected to an oil spill, in time they would be able to cleanse themselves and return to normal. Marine scientists at the Woods Hole Oceanographic Institution learned this was not necessarily true. The reason was a local oil spill that may have been a form of a blessing.

On September 16, 1969, the barge *Florida* ran aground and ruptured her steel hull, spilling 60,000 to 70,000 gallons of No. 2 fuel oil along the beaches of West Falmouth, Massachusetts. This is the kind of oil that is used in many homes for heating. The spill was 150 times smaller than that of the *Torrey Canyon.* The West Falmouth area, within 10 miles of the scientists at the oceanographic institution, offered an excellent opportunity to study the effects of an oil spill from the moment it happened and to study the long-term results, such as damage or recovery.

Observations made detected that the oil-soaked beaches were covered with dead or dying life of all kinds. There were fish, clams, oysters, scallops, crabs and worms. The tide pools were filled with marine worms, which were forced from their holes in the mud or sand and lay exposed, rotting under the sun. Later lobsters and some kinds of bottom fish, such as

One of the rigs used to clean up oil spills.
PHOTO: COASTAL SERVICES DIVISION, OCEAN WORLD CORPORATION

scup and tomcod, were washed up on the beaches. Because these animals live in deeper waters, it was learned that the oil affected not only the tidal waters but also the bottom below low tide.

A long-range study produced a startling discovery. After many months some of the polluted sea animals that lived during the West Falmouth oil spill were reexamined. Some oysters, which had been removed from the area and placed in clean water, retained as much oil in their systems as they had had six to eighteen months earlier. There was no change in the composition or quantity of the oil. It was then stated that when shellfish are so contaminated with oil, they are unable to cleanse themselves thoroughly of the pollution.

After the Woods Hole study the results of a comprehensive study of the Santa Barbara oil spill were released. The results of the twelve-month study revealed that the damage to beaches, plants and animals was less than predicted. The whole area was on its way to recovery, the report said. There were no effects of oil pollution on any of the zooplankton and phytoplankton. And the production of fish and larvae, as well as sea plants, did not suffer lasting ill effects. It was learned that one kind of barnacle was smothered and that 4,000 out of a total of 12,000 sea birds were killed. But migrating gray whales were not bothered.

The reason, according to the Santa Barbara report, that there was not a large mortality rate among animals is that toxins are the lightest components of oil and therefore float on the surface of the water and evaporate. Of course this disagrees with the Woods Hole report that stated that the toxins are the most persistent of the oil components. The one difference in the spills may explain the disagreement. The California oil spill occurred miles at sea, allowing the oil to drift at sea and speed up evaporation. The West Falmouth spill was close to shore, leaving no time for evaporation and thus more damage.

The presence, or absence, of an oily smell is not a good method of determining if a marine animal is polluted or not. Only a small part of any petroleum product has a strong odor and this may be lost while poisons are still inside the animal. Frying or boiling the shellfish may rid the animal of the oily smell, but in no way rids it of the poison.

The oil in the sea may also interfere with the normal day-to-day life of plants and animals. The changes in the chemistry could cause false responses in fish and other animals. The animals in the ocean depend on chemical attraction to find their food, to mate and to select the best places to live. With the changes in chemistry there may be enough interference to cause an animal to be unable to find food or to attract a mate during the mating season. Either way the potential of losing a species, or at least depleting the species, is an actual fact.

It has been estimated that by the year 1980 one-third of the world's oil production will come from the offshore coastal area. By the mid-1980s the production will have jumped to about half of the world's total. Ultimately the offshore wells over the continental shelves will produce 1,600 billion barrels of oil including natural gas. That is an alarming figure and poses the question: What will happen to the shores near this production?

That is why the lawmaker is caught in the middle when it comes to determining what should be done on the continental shelves. Should more areas be opened up for extracting oil, sand, gravel, tin, copper or whatever? Are conservationists exaggerating in their claim that we will all suffer greatly if the shelves are disturbed to any great degree? And is this a new problem that is only temporary and will go away, drift off with the winds of time?

George Vancouver, who explored and surveyed the north Pacific coast, wrote, "The surface of the sea, which was perfectly smooth and tranquil, was covered with a thick, slimy substance, which, when separated or disturbed by a little agitation, became very luminous, while the light breeze that came principally from the shore brought with it a strong smell of tar or some such resinous surface, which covered the ocean in all directions within the limits of our view, and indicated that in the neighborhood it was not subject to much agitation." The survey took three years to complete . . . during the years from 1792 to 1794!

It is not foolhardy to say that the continental shelves are of tremendous importance to man, so great in fact that without the shelves man would find it very difficult to survive on this planet, if indeed he could survive at all. It would only be foolhardy for a great many people to ignore the shelves, ignore what is there, the fish, the oil, the gas and the minerals. Everyone should take an interest because, as sure as we are all a part of mankind, we all depend on the shelves for something. And it may very well be for life itself.

Further Reading

Bates, Marston, *The Forest and the Sea*, New York, Random House, Inc., 1960

Buchsbaum, Ralph, *Animals Without Backbones*, Chicago, University of Chicago Press, 1948

Carson, Rachel, *The Sea Around Us*, New York, Western Publishing Company, Inc., 1958

Marx, Wesley, *The Frail Ocean*, New York, Coward-McCann, 1970

Pinney, Roy, *Underwater Archeology*, New York, Hawthorn Books, Inc., 1970

Ray, Carleton and Ciampi, Elgin, *Underwater Guide to Marine Life*, New York, A. S. Barnes and Company, Inc., 1956

Scheffer, Victor, *Year of the Whale*, New York, Charles Scribner's Sons, 1969

Schmitt, Waldo L., *Crustaceans*, Ann Arbor, University of Michigan Press, 1965

Waters, John F., *Marine Animal Collectors*, New York, Hastings House, Publishers, Inc., 1969

Waters, John F., *Some Mammals Live in the Sea*, New York, Dodd, Mead and Company, 1972

Waters, John F., *The Sea Farmers*, New York, Hastings House, Publishers, Inc., 1970

Index

About the Author

John F. Waters has been a bomb demolitions expert, a newspaperman and a fifth-grade school teacher. After graduating from the University of Massachusetts, with a bachelor's degree in education and science, he pursued his principal interests, marine science, oceanography, writing and beachcombing. This led him to become the author of over twenty books for both adults and young people dealing with various aspects of marine science.

Three of Mr. Waters's books have been chosen as Outstanding Science Books by the National Science Teachers Association and the Children's Book Council. Two of his books have been Junior Literary Guild selections.

John F. Waters, with his wife and four children, continues to beachcomb the sandy shores of Cape Cod.